"A marvellous little book, full of nuggets of wisdom from the 'who's who?' of our industry. I highly recommend this book to all young and aspiring geoscientists."

Dan Hampson
co-founder of Hampson–Russell

"This is a great book... The contributing authors are among the best known names in our profession. The subject each author selects is an essential 'thing' that we all need to know about geophysics. I predict that when you get a copy of this book in your hand, you will look at every page."

Bob A Hardage
President of SEG

"I was grinning to myself as I read some of the comments. I liked the informal tone and the down-to-earth advice. The bite-sized pieces of advice will be most useful to students starting out in the field. It's a fundamental truth that it is way more efficient to progress in your discipline if you gaze at the horizon standing on the shoulders of those who came before... This book should make a useful addition to any new geophysicist's library!"

Henry Posamentier
Seismic geomorphologist

"Fascinating. In the current world of instant gratification this provides rapid 'bites' of insight into many aspects of geophysics, seen through the eyes of some of the science's best practitioners."

David Monk
President-Elect of SEG

52 THINGS YOU SHOULD KNOW ABOUT GEOPHYSICS

EDITED BY MATT HALL & EVAN BIANCO

Agile Libre

First published in 2012 by Agile Libre
Nova Scotia, Canada. *www.agilelibre.com*

Technical editors Matt Hall & Evan Bianco • *Managing editor* Kara Turner
Designer Neil Meister, MeisterWorks • *Indexer* Linda Lefler
Cover design electr0nika • *Printing* Amazon CreateSpace

Library and Archives Canada Cataloguing in Publication

52 things you should know about geophysics / edited by Matt Hall and Evan Bianco
Includes bibliographical references and index.

ISBN 978-0-9879594-0-9

1. Geophysics. I. Hall, Matt, 1971- II. Bianco, Evan, 1982-
III. Title: Fifty-two things you should know about geophysics.

QC806.F54 2012 550 C2012-902408-2

Who we are

Agile Libre is a new, independent publisher of technical books about the subsurface. The book you are holding is its first book, but there will be more. We have a passion for sharing, so our books are openly licensed and inexpensive to buy.

Our aim is to be useful, relevant, and interesting. How can we make your life better? Send your ideas to *hello@agilelibre.com.*

Where to get this book

You will find this book for sale at *agilelibre.com*, and also at Amazon's various stores worldwide. Professors, chief geoscientists, managers, gift-givers: if you would like to buy more than ten copies, please contact us for a discount at *hello@agilelibre.com.*

About open licenses

Colophon

This book was compiled in Google Docs and Microsoft Word, and laid out on a Mac using Adobe InDesign with the MathMagic plug-in. The cover typeface is Avant Garde Gothic and the text typefaces are Minion and Myriad. The figures were prepared in Inkscape. It was published through Amazon's CreateSpace.

Contents

Alphabetical

Contents

By theme

Theme: FUNDAMENTALS

Tags: BASICS • MAPPING

CONCEPTS

GEOLOGY • ANALOGS

INTERPRETATION

POWERTOOLS

ATTRIBUTES • NINJA SKILLS

PRE-STACK

ROCK PHYSICS • PRE-STACK • PROCESSING

QUANTITATIVE

MATHEMATICS • ANALYSIS

INTEGRATION

TEAMWORK • WORKFLOW

INNOVATION

HISTORY • INNOVATION • TECHNOLOGY

SKILLS

LEARNING • CAREER • MANAGING

Introduction

This book is an experiment. From the minute we first talked about the idea in May 2011, we have only been sure about one thing: we aren't sure about anything. Well, okay, maybe a few things... We knew we wanted to bring together a group of personal heroes, mentors, friends, colleagues, and well-known interpretation visionaries. We knew we wanted short, highly relevant, fun-to-read, easy-to-share, and above all useful, essays. We knew we wanted to make a book that cost under $20, with an ebook version. We knew we wanted to make it easy to share. And we knew we wanted to do it ourselves.

Lots of people asked us, 'Why not ask the SEG to publish it?' A fair question, since they obviously know what they're doing and we, well, don't. But we're curious people, and maybe we have grand ideas about doing a series, and straying away from geophysics, and doing who-knows-what. We knew enough to know that we didn't know enough to give the project away. We wanted to own it, experiment with it, and see where it ended up.

It has not ended up anywhere yet, far from it, but this book is the end of the beginning. Now we know how to invite people, how to collate, read, edit, reread, re-edit, and proof 52 pieces of writing by 39 authors. We know how to design cover art, build page layouts, and typeset equations. And we know how to make a manuscript into an actual book that you can sell to actual people. We are not astute businessmen — we have no idea how many we will sell (100? 1000? 10 000?), or if the project will come close to breaking even. We hope it does, not because we want to make lots of money, but because we want to do it again.

It all started at the annual convention of the Canadian Society of Exploration Geophysicists in May 2011. We'd already decided that for the book to be remarkable, we'd need to somehow convince one or two of a handful of highly accomplished people to contribute something. We weren't at all sure how this bit would work, if anyone would even listen. Evan mentioned the project to Brian Russell at the convention, and he seemed genuinely interested, enthused even. Encouraged enough to put some invites out, we emailed a few potential authors a couple of weeks later.

Brian — Brian Russell, über-geophysicist, ex-president of SEG and CSEG, award-winning entrepreneur — emailed his first contribution, *Don't neglect your math*, 23 hours later. It was perfect, exactly the sort of thing we wanted. We were ecstatic. And we knew the book was going to work. When Matt expressed his surprise at getting such a fast response from such an awesome geoscientist, his wife Kara (and the book's managing editor) was incredulous: 'It's not surprising at all: that's precisely why he's awesome,' she said.

Well, we had many more moments like this as contributions came in from the amazing, and incredibly busy, geoscientists we'd written to. We don't want to make it sound like this project is all about celebrities, far from it. Admittedly, getting chapters early on from well-known veterans was highly motivating, but ultimately no more so than the insightful chapters from the many energetic young professionals who also contributed.

One of our core principles at Agile is that expertise is distributed — we all have specializations and experience that others can enjoy hearing about and learning from. We are committed to this idea. Insight and creativity come from everyone in this industry, not just from veterans, chiefs, and superstars. If we can cultivate a culture of more open and accessible writing and knowledge, we believe we can spark more ideas, connect more people, and have more fun than ever.

Welcome to the first of these short collections of *Things You Should Know*. If you like it, please share it — give this book to someone you respect.

Matt Hall & Evan Bianco
Nova Scotia, May 2012

Anisotropy is not going away

Vladimir Grechka

One of the first things geophysics students learn in an introductory seismology course is the wave equation. It is usually introduced under an explicitly stated or tacit assumption of isotropy. This assumption is certainly useful because it allows the students to harness their intuition and even rely on their childhood experience, from which they remember that rocks thrown in water make circles on its surface. Those circles, expanding as the waves propagate, are the wavefronts described by the wave equation; they have circular shapes exactly because water is homogeneous and isotropic.

The properties of homogeneity and heterogeneity are easy to grasp because we readily observe them every day. For instance, while pure water is normally perceived as homogeneous, a hearty stew would be an example of a heterogeneous substance. Physical properties vary spatially in such substances. As the seismology course proceeds, students are introduced to another property of solids — anisotropy. In an anisotropic substance, a measured quantity such as the wave propagation velocity depends on the direction in which it is measured rather than the spatial location of the measurement. Anisotropy is not as readily available to our casual observation as heterogeneity; after all, we see circles and not ovals when we throw rocks in water.

The understanding of anisotropy is critically important for exploration geophysics because the majority of sedimentary formations — through which seismic waves propagate on their way to and from hydrocarbon reservoirs — are shales. They consist of highly anisotropic clay particles, which make seismic anisotropy of shales routinely observable in exploration practice. There are other reasons for the subsurface anisotropy too, notably fractures and non-hydrostatic stresses whose abundant presence is widely documented in the geophysical and geological literature. Ignoring seismic anisotropy, as our industry often does, causes various issues in seismic data processing, some of which might entail substantial financial losses. Among them are:

- Mispositioned and blurred seismic images, which could lead to missed exploration targets.
- Improperly placed or poorly imaged faults, which sometimes result in unexpected and expensive-to-fix drilling problems.

The advice is straightforward: understand the theory and learn from the practice.

- The inability to extract quality information from converted-wave and shear-wave data.

On top of these things, certain other tasks, such as seismic characterization of fractures, simply cannot be implemented with isotropic assumptions because seismic anisotropy is the very reason for the observed signatures.

As practitioners, what should we do? The advice is straightforward: understand the theory and learn from the practice. There are numerous papers and a few good books that can form a paradigm for applications of seismic anisotropy in exploration and development settings. Embracing such a paradigm is especially important for young geophysicists because, as our data-acquisition technologies improve, we will see more rather than less seismic anisotropy and, as our data-processing and interpretation methods mature, we should be able to relate the estimated anisotropy to fluids in the rock and the sub-wavelength rock fabric more and more precisely.

Beware the interpretation–to–data trap

Evan Bianco

Some say there is an art to seismic interpretation. Art is often described as any work that is difficult for someone else to reproduce. It contains an inherent tie to the creator. In this view, it is more correct to say that seismic interpretation is art.

Subsurface geoscience in general, and seismic interpretation in particular, presents the challenge of having long, complex workflows with many interim sections, maps, and so on. As the adage of treasure and trash goes; one person's interpretation is another person's data. The routine assignment of interpretations as data is what I call the interpretation–to–data trap, and we all fall into it.

Is a horizon an interpretation, or is it data? It depends on whom you ask. To the interpreter, their horizons are proud pieces of art, the result of repetitious decision making, intellectual labour, and creativity. But when this art is transferred to a geomodeller, for instance, it instantly loses its rich, subjective history. It becomes just data. Without fail, interpretations become data in the possession of anyone other than the creator. It is a subtle but significant concept. And consider the source from which the interpreters' horizon manifests: seismic amplitudes. Stacked seismic data is but one solution from a choice of migration algorithms, which is itself an interpretive process. More data from art. To some extent, this is true for anything in the subsurface, whether it be the wireline log, production log, or well top.

There are a number of personal and social forces that deepen the trap, such as lack of ownership or lack of foresight. Disowning or disliking data is easy because it is impersonal; disowning your own interpretation however is self-sabotage. People are rarely blamed for spending time on something within their job description, but it is seldom anyone's job to transfer the implicit (the assumptions, the subjectivity, the guesswork, the art) along with the explicit. It takes foresight and communication at a personal level, and it takes a change in the culture of responsibilities, and maybe even a loosening of the goals on a team level.

Because of the interpretation–to–data trap, we must humbly recognize that even the most rigorous seismic interpretation can be misleading as it is passed down-

The routine assignment of interpretations as data is what I call the interpretation–to–data trap, and we all fall into it.

stream and farther afield. If you have ever felt that a dataset is being pushed beyond its limits, that a horizon was not picked with sufficient precision for horizontally steering a drill bit, or a log-conditioning exercise was not suitable for an AVO analysis, it was probably a case of interpretation being mistaken for data. Sometimes these are inevitable assumptions, but it doesn't absolve us of responsibility.

I think there are three things you can do. One is for giving, one is for receiving, and one is for finishing. When you are giving, arm yourself with the knowledge that your interpretations will transform to data in remote parts of the subsurface realm. When receiving, ask yourself, 'is there art going on here?' If so, recognize that you are at risk of falling into the interpretation–to–data trap. Finally, declare your work to be work in progress, not because it is a way to cop-out or delay finishing, but because it is an opportunity to embrace iteration and make it a necessary part of your team dynamic.

Calibrate your intuition

Taras Gerya

I am a geodynamicist. Geodynamics aims to understand the evolution of the earth's interior and surface over time. The following simple exercise explores the subject of geodynamics in the context of the availability of data. There are many parallels with modelling systems in exploration geophysics.

Let's imagine an ideally symmetrical earth with physical properties (density, viscosity, temperature, etc.) as functions of depth and time. A simple two-dimensional time–depth diagram covering the earth's entire history and its interior will thus be a schematic representation for the subject of geodynamics. For the real earth such a diagram should be four-dimensional, but this rectangular diagram will do for us. The entire diagram should be then covered by data points characterizing the physical state of the earth at different depths, ranging from 0 to 6000 km, and for different moments of geological time, ranging from 0 to around 4.5 billion years ago. However, the unfortunate fact for geodynamics is that observations for such systematic coverage are only available along the two axes of the diagram: geophysical data for the present-day earth structure and historical record in rocks formed close (typically within a few tens of kilometres) to the earth's surface. The rest of the diagram is thus fundamentally devoid of observational data, so we have to rely on something else.

What else can we use? Scientific intuition based on geological experience and modelling based on fundamental laws of continuum mechanics! However, our intuition cannot always be suitable for geodynamical processes that are completely out of human scales in time (a few years) and space (a few metres). We have to accept that some of these processes could look completely counter-intuitive to us. For example, acceptance of plate tectonics was delayed by relying on commonsense logic: a solid mantle conducting shear seismic waves can only deform elastically not allowing for the movement of continents over thousands of kilometres. The crucial point that was finally understood by the scientific community is that both viscous (fluid-like) and elastic (solid-like) behaviour are characteristic of the earth, depending on the time scale of deformation. The mantle, which is elastic on a human time scale, is viscous on geological time scales (>10 000 years) and can be strongly internally deformed by very slow solid-state creep.

The systematic use of both analog and numerical modelling is crucial to develop, test, and quantify geodynamic hypotheses.

The ways in which various geodynamic processes interact with each other can also be very difficult to conceive using only scientific intuition. This is why intuition in geodynamics should be — must be — assisted by modelling. In a way, modelling helps train our intuition for very deep and very slow geological processes that cannot be observed directly.

Another role of modelling is the quantification of geodynamic processes based on the sparse array of available observations. Consequently, the systematic use of both analog and numerical modelling is crucial to develop, test, and quantify geodynamic hypotheses — and perhaps most questions about the earth.

Don't ignore seismic attenuation

Carl Reine

In general terms, seismic attenuation is the loss of elastic energy contained in a seismic wave, occurring through either anelastic or elastic behaviour. Anelastic loss, or *intrinsic attenuation*, is a result of the properties of the propagation medium. It causes a fraction of the wave's energy to be converted into other forms such as heat or fluid motion. Non-intrinsic effects like multiple scattering are collected under the term *apparent attenuation*; they are so many and varied that intrinsic attenuation is difficult to isolate.

Attenuation of a seismic wave results in amplitude loss, phase shifts due to the associated dispersion, and the loss of resolution over time. This makes interpretation of the stacked data more difficult, and introduces an amplitude gradient in pre-stack data that is not predicted by reflectivity alone. These problems can be mitigated by the use of an inverse Q-filter which attempts to reverse the attenuation effects based on a measured attenuation field. Measures of attenuation are useful in their own right, because through them we can infer reservoir properties for characterization purposes.

The mechanisms of intrinsic attenuation have been referred to as either jostling or sloshing losses, relating to losses from dry-frame or fluid-solid interactions respectively. Jostling effects involve the deformation or relative movement of the rock matrix, for example due to friction from intergranular motion or the relative motion of crack faces. More significant in magnitude are the sloshing effects, which occur when pore fluids move relative to the rock frame. These fluid effects are characterized by the scale over which pressure is equalized, from large-scale Biot flow, between the rock frame and the fluid, to squirt flow caused by the compression of cracks and pores.

Attenuation may be described as the exponential decay of a seismic wave from an initial amplitude A_0 to an attenuated state A over distance z, quantified by an attenuation coefficient α:

$$A = A_0 e^{-\alpha z}$$

Various models describe the behaviour of α with frequency, and the relationship between the frequency-dependent velocity $V(f)$ and the quality factor Q. Although many of the intrinsic attenuation mechanisms, specifically the

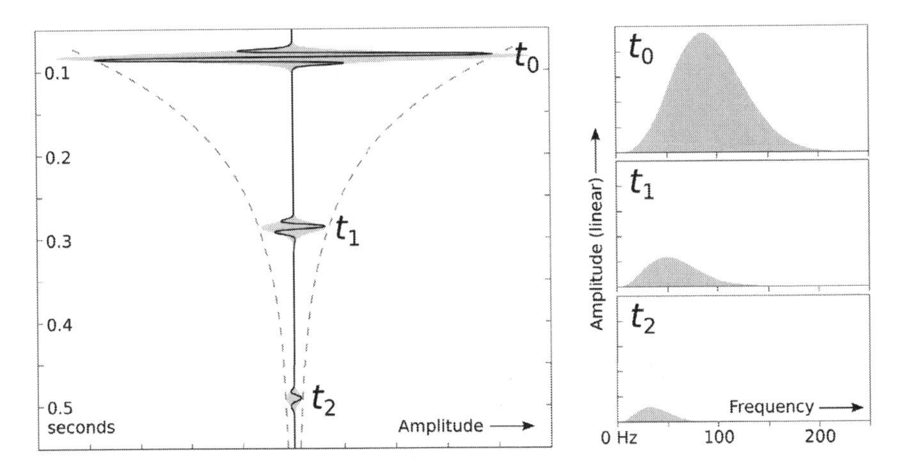

sloshing losses, are dependent on frequency, Q is often assumed to be constant over the limited bandwidth of seismic experiments, resulting in α being linear with respect to frequency:

$$\alpha = \pi f / VQ$$

Q may also be defined by the complex behaviour of the elastic modulus M:

$$Q = \mathrm{Re}(M) / \mathrm{Im}(M)$$

Seismic attenuation may be measured from different experimental configurations and at different frequency scales. The general method is to analyse the change in amplitude or frequency content or both, with propagation distance. For exploration seismic, two of the more common approaches are:

1. **Spectral ratio** — The spectrum is observed at two points along its travel-path, $S_1(f)$ and $S_2(f)$, with a time separation Δt. There is an inverse linear relationship between Q and the natural logarithm of the ratio of S_2 to S_1:

$$\ln\left[\frac{S_2(f)}{S_1(f)}\right] = -\frac{\pi \Delta t f}{Q} + \ln(PG)$$

   The intercept term is determined by frequency independent amplitude effects such as energy partitioning P and geometric spreading G.

2. **Frequency shift** — In which the shift of the centre frequency of the spectrum is related to attenuation of a modelled spectrum.

Seismic attenuation is rarely estimated — it is an often overlooked attribute. But the desire for quantitative geophysical attributes and improved data quality should win it more attention, especially as high-fidelity data allow ever more quantitative analysis. The effects of attenuation can no longer be ignored.

Don't neglect your math

Brian Russell

Doing mathematics is like exercising. Do a little bit every day and you stay in shape, either intellectually (in the case of math) or physically (in the case of exercising). Neglect it and your muscles (intellectual or physical) fade away.

Geophysics is a hard science. By that I mean that it is a science based on 'hard' facts, but also that it can be difficult. We all struggled through tough math and physics classes at university to get our degrees. But once we were in the working world, especially if we became seismic interpreters, we tended to leave the details to the specialists. Indeed, picking up a copy of *Geophysics* and trying to read every article is a daunting task. And I do not expect that every exploration geophysicist should be able to understand the latest implementation of Green's functions in anisotropic depth imaging. However, I do think that an appreciation of some of the fundamental applied mathematical ideas in our profession can go a long way towards enhancing your enjoyment and appreciation of your day-to-day job.

Two examples

Let me illustrate my point with two equations. Let us start with:

$$d = Gm$$

where d is our data, a set of n geophysical observations, m is our model, a set of k model parameters, and G is a linearized relationship that relates the observations to the parameters. This ubiquitous equation can be found in every area of geophysics, from seismology through potential fields to electromagnetic theory. The simplicity of the way I have written the equation hides the fact that d is usually written as an n-dimensional vector, m as a k-dimensional vector, and G as an n row by k column matrix.

Solving the equation is a little more difficult. Since n is usually greater than k, the solution can be written:

$$m = (G^T G + \lambda I)^{-1} G^T d = C^{-1} h$$

where C is the autocorrelation matrix found by multiplying the G matrix by its transpose G^T (and adding a little pre-whitening by multiplying the value λ

The way that you react to these equations tells me a lot about you as a geophysicist.

by I, the k by k identity matrix), and h is the zero-lag cross-correlation vector, found by multiplying the transpose of the G matrix by the data vector. Again, this equation, sometimes called the Normal Equation, is ubiquitous in geophysics. It is the basis of deconvolution, AVO attribute analysis, post- and pre-stack inversion, refraction and reflection statics, and so on. So, what lesson should we take away from these equations?

My advice

The way that you react to these equations tells me a lot about you as a geophysicist. If you are thinking: 'what's the big deal, I use those types of equations every day,' you probably don't need my advice. If you are thinking: 'yes, I saw those equations once in a class, but haven't thought about them for years,' perhaps I can inspire you to look at them again. On the other hand, if you are thinking: 'why would I ever need to use those boring-looking equations,' you are a tougher challenge! I would recommend starting with these equations and really trying to understand them (perhaps you will need to dust off your linear algebra, and I recommend the book by Gilbert Strang). Then, pick up a copy of *Geophysics*, or any geophysics textbook, and see how many of the equations can be expressed in the same way. Or, take some quantitative industry training courses and see what the mathematics is really telling you about your data.

I guarantee it will be good for you!

References

Strang, G (2009). *Introduction to Linear Algebra*. Wellesley–Cambridge Press, 574 pages; *math.mit.edu/linearalgebra*

Don't rely on preconceived notions

Eric Andersen

I am amazed that geophysics works. When you think about the seismic experiment, there are so many assumptions and technologies involved. Couple this with the fact that the earth is a filter with endless complexities and non-unique characteristics, and it is incredible that we find hydrocarbons at all.

However, the days of finding large reserves under simple bumps are gone. Today's exploration plays are plagued with intense structures, stratigraphic variability, anisotropy, high pressures, low permeabilities, and deep targets. The geophysical problem is ever more complicated.

Traditional interpretation techniques are insufficient in these areas. It's not about mapping the wiggle, it's about what causes the wiggle to look that way. It's about rock physics. A seismic signal has three properties: amplitude, phase, and frequency. Seismic attribute analysis is an evaluation of one or more of these properties in some detail.

The tricky thing with seismic attribute analysis is that it is still not well understood by most explorationists. Perceptions about the techniques involved are typically based on old technologies, tales of failure, or conjecture.

It is interesting to me that seismic interpreters routinely pick seismic wiggles and use these correlations to confidently map structures and reservoirs. However, when the same data are inverted and displayed in colour, hands go up and cries of 'non-uniqueness' fill the room. Let's be clear: seismic is non-unique. Many interpretations can be made on the same dataset without proper controls.

Generating and interpreting attributes is similar to solving a mathematical problem. If you follow the general rules and make valid assumptions, you will arrive at a realistic solution. The concept is not unlike learning mathematics in school. Remember BEDMAS, a method taught for solving algebraic equations? BEDMAS is an acronym to help remember the order of operations used for solving these equations: Brackets, Exponents, Division, Multiplication, Addition, and Subtraction. Unfamiliar and difficult problems can be solved if you follow the guidelines. However, if you deviate from the workflow, skip a step, or make a poor assumption, the result is adversely affected.

When using seismic attributes, don't rely on historical or preconceived notions.

Failure in seismic attribute problems occurs for similar reasons. Skipped steps, poor assumptions, lack of diligence in processing or well correlations, and ineffective conditioning of input data all contribute to unsatisfactory results.

It is also important to communicate the results effectively. Geologists discuss a formation's porosity as a measured entity. Geophysicists may derive a porosity using seismic attributes, but it will likely be over the thickness of the reservoir. The two will not be exactly the same, and they should not be presented as the same. Generally one is specific, the other is relative. You need to know what you are solving for.

The learning you should take from this is: When using seismic attributes, don't rely on historical or preconceived notions. Evaluate the problem for yourself, follow the proper rules with diligence, and present the results in a way people understand. If you do this, you will find that geophysics works and can be used to solve the most difficult of problems.

Evolutionary understanding is the key to interpretation

Clare Bond

If we consider the maximum number of possible interpretations to a seismic image, given the conceptual uncertainty of the data (see the chapter *Recognize conceptual uncertainty and bias*), there is a question of how best to narrow down the interpretational possibilities. One method would be to assign each possible concept with a probability, but given the possible permutations and inherent bias already in the data from acquisition and processing, this method raises further questions of how best to assign probabilities. It's also likely to generate the same, or most likely, solution and sometimes we might want to think outside the box.

Experiments investigating seismic interpretation have shown that particular techniques have a strong influence on an interpreter's ability to get the correct answer in model data. Notable amongst the techniques is evidence of genetic or *evolutionary* thought processes. Evidence of these evolutionary thought processes include drawing cartoons of, writing about, or annotating elements of the sequential evolution of the imaged geology. In fact it doesn't matter how the interpreter evidences the thought processes, only that they have checked their interpretation is genetically or evolutionarily viable. The use of evolutionary sketches or text forces an interpreter to reconstruct, even in a simple way, the geometric evolution of the geology over time. By showing how the geometries evolve, an interpreter demonstrates that the final geometry in the interpreted model is valid and that the interpretation is plausible. For structural geologists this process is more formally known as cross-section balancing or restoration and is used to validate the geometrical arrangement of picked faults and horizons. But the method works for other disciplines, too.

In an experiment investigating how *experts* (in this case defined as structural geologists) interpreted a structurally complex synthetic seismic image, we found that only 10 percent showed evidence of having thought about the structural evolution of the imaged geology. Those that did were three times more likely to get the correct interpretation (Bond et al. 2012). The few participants that used the technique, but failed to get the correct answer, proved in their own evolutionary sketches that their final interpretations were incorrect, as they were unable to generate the final geometry.

The message is really to make sure that the picks you have made make sense.

The message is really to make sure that the picks you have made make sense. It is all very well to focus on the data when interpreting, but the interpretation has to be geologically plausible. It would seem that simple validation techniques are infrequently used and they may well be the key to understanding which of the possible interpretations of a dataset are plausible. So next time you are stuck on an interpretation, grab the coloured pencils.

References

Bond, C E, R J Lunn, Z K Shipton, and A D Lunn (2012). What makes an expert effective at interpreting seismic images? *Geology* **40**, 1, 75–78.

Explore the azimuths

David Gray

We should process 3D seismic using 3D concepts. This means accounting for and using azimuthal variations in the seismic response (e.g. Gray et al. 2009). Recent results from azimuthal AVO analysis (e.g. Gray and Todorovic-Marinic 2004) and shear-wave birefringence (Bale 2009) have shown that there is significant variation in azimuthal properties over small areas. The implication is that local structural effects, like faults and anticlines, dominate over regional tectonic stresses in azimuthal seismic responses. It is possible that processing algorithms that remove average properties, like surface-consistent methods, may dampen regional effects relative to local effects, but as far as I am aware this concept remains untested at this time. Regardless, there is an imprint of local structural effects on the azimuthal properties, probably caused by the opening and closing of pre-existing fractures in the rock by these local structures.

The largest azimuthal effects come from the near-surface. Examination of residual azimuthal NMO above the oil sands of Alberta have revealed up to 15 ms of residual moveout at depths of less than 200 m (e.g. Gray 2011) and more than 20 ms of birefringence at similar depths (Whale et al. 2009). There is currently some discussion as to why this apparent anisotropy is observed so shallow in the section. Various explanations include stress, fractures, heterogeneity along different azimuthal ray paths, surface topography, and so on. Regardless of their source, these effects propagate all the way down through the data and affect the ability to properly process the data and estimate amplitude attributes.

Azimuthal effects are not restricted to land data. Significant azimuthal effects have been observed in narrow-azimuth towed-streamer seismic data (e.g. Wombell et al. 2006). Application of azimuthal NMO to this seismic volume results in much better offset stacks and significant reduction of striping in timeslices.

The above discussion focuses on the use of azimuthal — that is, 3D — NMO to improve the processing of 3D seismic volumes. This tool is readily available and relatively easy to use. There are other applications where the use of 3D azimuthal concepts and the understanding that properties do vary with azimuth should help to improve the seismic image:

The largest azimuthal effects come from the near-surface... and propagate all the way down through the data.

- Azimuthal migration (Gray and Wang 2009) with azimuthal velocities (e.g. Calvert et al. 2008);
- Incorporating local azimuthal variations into surface-consistent algorithms such as deconvolution, scaling, and statics;
- Amplitude inversion for elastic properties (e.g. Downton and Roure 2010), noise attenuation, etc.

References

Bale, R (2009). Shear wave splitting applications for fracture analysis and improved imaging: some onshore examples, *First Break* **27** (9), 73–83.

Calvert, A, E Jenner, R Jefferson, R Bloor, N Adams, R Ramkhelawan, and C St. Clair (2008). Preserving azimuthal velocity information: Experiences with cross-spread noise attenuation and offset vector tile preSTM, SEG *Expanded Abstracts* **27**, 207–211

Downton, J and B Roure (2010). Azimuthal simultaneous elastic inversion for fracture detection, SEG *Expanded Abstracts*, **29**, 263–267

Gray, D and D Todorovic-Marinic (2004). Fracture detection using 3D azimuthal AVO. CSEG *Recorder* **29** (10).

Gray, D, D Schmidt, N Nagarajappa, C Ursenbach, and J Downton (2009). An azimuthal-AVO-compliant 3D land seismic processing flow. CSPG–CSEG–CWLS *Expanded Abstracts*.

Gray, D and S Wang (2009). Towards an optimal workflow for azimuthal AVO. CSPG–CSEG–CWLS *Expanded Abstracts*.

Gray, D (2011). Oil sands: not your average seismic data. CSPG–CSEG–CWLS *Expanded Abstracts*.

Whale, R, R Bale, K Poplavskii, K Douglas, X Li, and C Slind (2009). Estimating and compensating for anisotropy observed in PS data for a heavy oil reservoir. SEG *Expanded Abstracts* **28**, 1212–16.

Wombell, R (2006). Characteristics of azimuthal anisotropy in narrow azimuth marine streamer data. EAGE *Expanded Abstracts* **68**.

Five things I wish I'd known

Matt Hall

For years I struggled under some misconceptions about scientific careers and professionalism. Maybe I'm not particularly enlightened, and haven't really woken up to them yet, and it's all obvious to everyone else, but just in case I am, I have, and it's not, here are five things I wish I'd known at the start of my career.

Always go the extra inch. You don't need to go the extra mile — there often isn't time and there's a risk that no one will notice anyway. An inch is almost always enough. When you do something, like work for someone or give a presentation, people only really remember two things: the best thing you did, and the last thing you did. So make sure those are awesome. It helps to do something unexpected, or something no one has seen before. It is not as hard as you'd think — read a little around the edges of your subject and you'll find something. Which brings me to…

Read, listen, and learn. Subscribe to some periodicals, preferably ones you will actually enjoy reading. You can see my favourites at *ageo.co/IsdDxB*. Go to talks and conferences, as often as you reasonably can. But, and this is critical, don't just go — take part. Write notes, ask questions, talk to presenters, discuss with others afterwards. And learn: do take courses, but choose them wisely. In my experience, most courses are not memorable or especially effective. So ask for recommendations from your colleagues, and make sure there is plenty of hands-on interaction in the course, preferably on computers or in the field. Good: Dan Hampson talking you through AVO analysis on real data. Bad: sitting in a classroom watching someone derive equations.

Write, talk, and teach. The complement to read, listen, and learn. It's never too early in your career to start — don't fall into the trap of thinking no one will be interested in what you do, or that you have nothing to share. Even new graduates have something in their experience that nobody else has. Technical conference organizers are desperate for stories from the trenches, to dilute the blue-sky research and pseudo-marketing that most conferences are saturated with. Volunteer to help with courses. Organize workshops and lunch-and-learns. Write articles for the *Recorder*, *First Break*, or *The Leading Edge*. Be part of your science! You'll grow from the experience, and it will help you to…

By far the best way to network is to help people. Help people often, for free, and for fun.

Network, inside and outside your organization. Networking is almost a dirty word to some people, but it doesn't mean taking people to hockey games or connecting with them on LinkedIn. By far the best way to network is to help people. Help people often, for free, and for fun, and it will make you memorable and get you connected. And it's easy: at least 50 percent of the time, the person just needs a sounding board and they quickly solve their own problem. The rest of the time, chances are good that you can help, or know someone who can. Thanks to the Matthew Effect, whereby the rich get richer, your network can grow exponentially this way. And one thing is certain in this business: one day you will need your network.

Learn to program. You don't need to turn yourself into a programmer, but my greatest regret of my first five years out of university is that I didn't learn to read, re-use, and write code. Read *Learn to program* to find out why, and how.

Geology comes first

Chris Jackson

A new, high-quality, 3D seismic reflection survey arrives in the office on a portable drive, or is transferred from a service company via FTP. The big question is, 'who should interpret the shiny new volume?'

Some companies believe that a geophysicist should interpret seismic data; seismic data is, after all, a geophysical data type, and geophysicists understand how such data are acquired and processed. They will understand the procedures used to depth-migrate the dataset and, perhaps, the steps that were undertaken to preserve amplitudes, correct statics, or suppress multiples. A geophysicist, however, may have no or only limited experience of the geological complexity that is present in the earth's subsurface. They may never — or only rarely — have been into the field to look at fluvial channels, segmented normal faults, or carbonate build-ups.

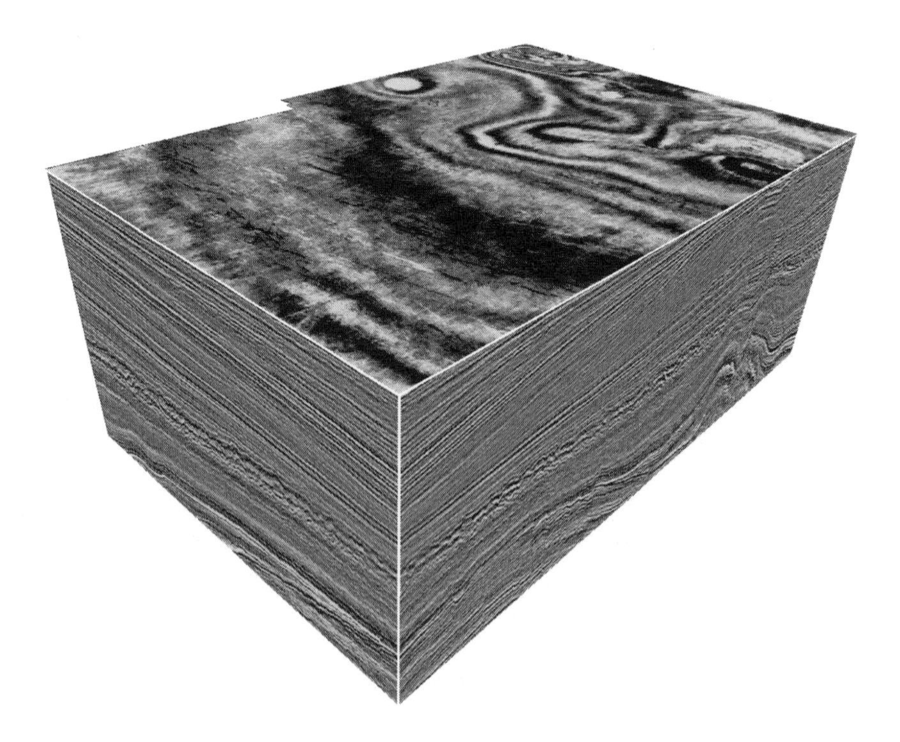

The big question is, 'who should interpret the shiny new volume?'

How can a geophysicist therefore make sense of the spatially complex and typically subtle features that are visible in many modern 3D seismic reflection datasets?

Perhaps the solution to the problem is to let a geologist, with a robust understanding of both structural geology and stratigraphy, but with some training in geophysics, interpret the data. They will understand the styles of secondary deformation that may be associated with a series of passive margin toe-thrusts. They will understand how seismic-scale clinoforms can provide insights into palaeo-sea-level changes, and the implications of this for identifying intervals that may be associated with lowstand deep-water reservoirs.

Geologists may, of course, forget that they are looking at an acoustic image of the earth rather than an outcrop, and they may be prone to interpreting too much geology. But surely it makes sense to have this new 3D seismic volume interpreted by the person with the best three-dimensional understanding of the complex structure and stratigraphy that exists in the earth's subsurface?

Acknowledgments

Offshore Netherlands F3 dataset from the dGB Earth Sciences Open Seismic Repository, *opendtect.org/osr*.

Geophysics is all around

José M Carcione

Hold a well-filled cup of milky coffee on a sunny day and on the liquid's surface you will see a *catacaustic*. This word has a Greek root and means 'burning curve'. The sun is a point source at infinity whose parallel rays hit the cup according to the laws of geometrical optics. The rays are reflected from the reflective inner wall of the cup generating the bright curve, the caustic, formed by the envelope of the rays. The cusp at the centre of the caustic is called the *paraxial focus* and the liquid surface is brighter below the caustic curve. This particular shape of caustic is called *nephroid*, meaning kidney-shaped.

The phenomenon was known to Huygens in about 1679, and Bernoulli described it mathematically as an *epicycloid*. But almost two centuries earlier, Leonardo da Vinci observed the caustic when he was experimenting with using concave mirrors to generate heat — he called them fire mirrors. He argued that given equal diameters, the one with a shallower curve concentrates the most reflected rays, and 'as a consequence, it will kindle a fire with greater rapidity and force'.

Syncline-shaped reflectors generate seismic reflections resembling these types of caustics.

The shallower curve...
will kindle a fire with greater
rapidity and force.

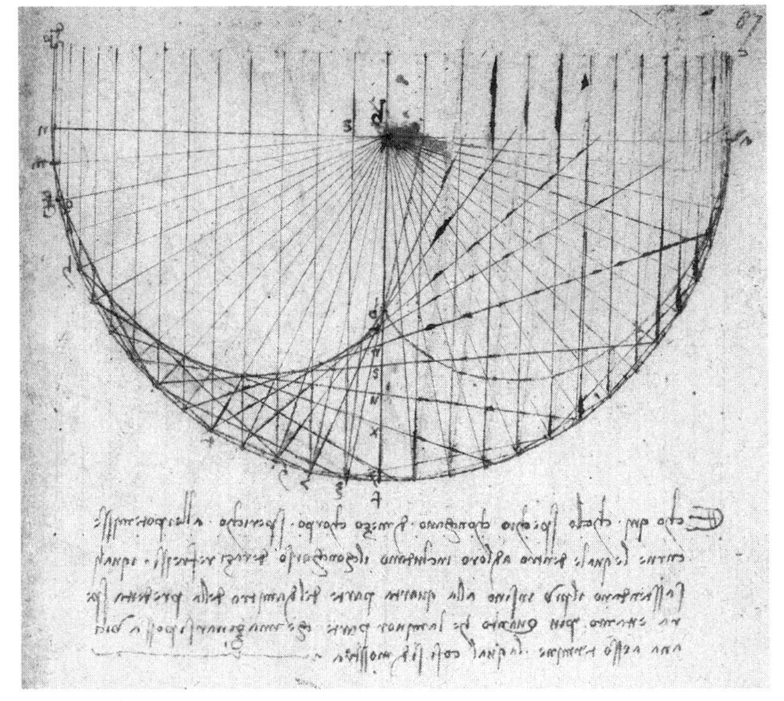

Leonardo da Vinci's early 16th-century sketch of a caustic, including his famous mirror writing.

Acknowledgments

This chapter is adapted from Carcione, J M (2007). *Wave fields in Real Media*, Elsevier. The cup image is original. The drawing is © The British Library Board, from *Codex Arundel*, MS 263, ff. 86v–87, ca. 1503–05, and used here with permission.

How to assess a colourmap

Matteo Niccoli

Seismic interpreters use colourmaps to display, among other things, time structure maps and amplitude maps. In both cases the distance between data points is constant, so faithful representation of the data requires colourmaps with constant perceptual distance between points on the scale. However, with the exception of greyscale, the majority of colourmaps are not perceptual in this sense. Typically they are simple linear interpolations between pure hue colours in red–green–blue (RGB) or hue–saturation–lightness (HSL) space, like the red–white–blue often used for seismic amplitude, and the spectrum for structure. Welland et al. (2006) showed that what is linear in HSL space is not linear in psychological space and that remapping the red–white–blue colourmap to psychological space allows the discrimination of more subtle variation in the data. Their paper does not say what psychological colour space is but I suspect it is CIE $L^*a^*b^*$. In this essay, I analyse the spectrum colourmap by graphing the lightness L^* (the quantitative axis of $L^*a^*b^*$ space) associated with each colour.

In this essay x is the sample number and y is lightness. The figure is shown here in greyscale, but you can view it in colour at *ageo.co/HLOS3a*. In the graph the value of L^* varies with the colour of each sample in the spectrum, and the line is coloured accordingly. This plot highlights the many issues with the spectrum colourmap. Firstly, the change in lightness is not monotonic. For example it increases from black ($L^*= 0$) to magenta M then drops from magenta to blue B, then increases again, and so on. This is troublesome if the spectrum is used to map elevation because it will interfere with the correct perception of relief, especially if shading is added. The second problem is that the curve gradient changes many times, indicating a non-uniform perceptual distance between samples. There are also plateaus of nearly flat L^*, creating bands of constant tone, for example between cyan C and green G.

Let's use a spectrum to display the Great Pyramid of Giza as a test surface (the scale is in feet). Because pyramids have almost monotonically increasing elevation there should be no substantial discontinuities in the surface if the colourmap is perceptual. My expectation was that instead the spectrum would introduce artificial discontinuities, and this exercise proved that it does.

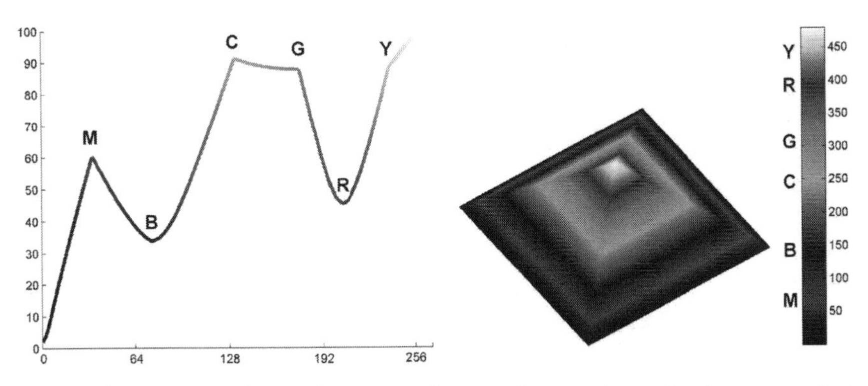

In an effort to provide an alternative I created a number of colourmaps that are more perceptually balanced. I will post all the research details and make the colourmaps available at *ageo.co/Jcgqgq*. The one used below was generated starting with RGB triplets for magenta, blue, cyan, green, and yellow (no red), which were converted to $L^*a^*b^*$. I replaced L^* with an approximately cube law L^* function, shown in the bottom left figure — this is consistent with Stevens' power law of perception (Stevens 1957). I then adjusted a^* and b^* values, picking from $L^*a^*b^*$ charts, and reconverted to RGB. The results are very good: using this colourmap the pyramid surface is smoothly coloured, without any perceptual artifact.

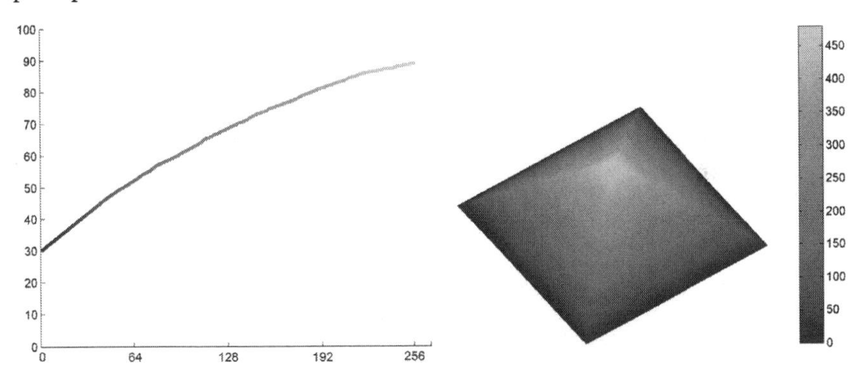

References

Stevens, S (1957). On the psychophysical law. *Psychological Review* **64** (3), 153–181. PubMed 1344/853

Welland M, N Donnelly (2006). Are we properly using our brains in seismic interpretation? *The Leading Edge* **25**,142–144.

Know your processing flow

Duncan Emsley

As computer technology has advanced, so has the complexity of our signal processing, noise attenuation, travel-time correction, and migration. So before embarking on any interpretation or analysis project involving seismic data, it is important to assess the quality and thoroughness of the processing flow lying behind the stack volumes, angle stacks, gathers or whatever our starting point is.

At the outset of 3D seismic surveying — and my career — about 25 years ago, computers were small. Migrations would be performed first in the inline direction in small swaths, ordered into crossline 'miniswaths', then miniswaths were stuck together, migrated in the crossline direction, then sorted back into the inline direction for post-migration processing. The whole process would take maybe two to three months — and that was just for post-stack migration. As computers evolved so did the migration flow, through partial pre-stack depth migrations in the days of DMO (dip moveout), then to our first full pre-stack time migrations. These days, complex pre-stack time or depth migrations can be run in a few days or hours and we can have migrated gathers available with a whole suite of enhancement tools applied in the time it took to just do the 'simple' migration part.

So when starting a new interpretation project, a first question might be: what is the vintage of the dataset or reprocessing that we are using? Anything more than a few years old might well be compromised by a limited processing flow, to say nothing of the acquisition parameters that might have been used. Recorded fold and offset have also advanced hugely with technology.

For simple post-stack migrated datasets, the scope of any interpretation might be limited to picking a set of time horizons. Simple processing flows like this often involve automatic gain control (usually just called AGC), which will remove any true amplitude information from the interpretation — the brightest bright spots might survive but more subtle relationships will be gone. Quantitative interpretation of any kind is risky.

In the pre-stack analysis world, optimally flattened gathers are critical. The flattening of gathers is a highly involved processing flow in itself, often involving several iterations of migration velocity model building, a full pre-stack time migration, and repicking of imaging, focusing, and stacking velocities. What happens in the event of class 1 or 2p AVO? Both involve phase reversal with offset and might require the picking of a semblance minimum in our velocity analysis. Is anisotropy present? How shall we correct for that? If anisotropy is present, we can build a simplified eta field into the migration, but is that too simple? We should scan for eta at the same time as we correct for NMO (normal moveout), but that is tricky. Eta is often picked after the NMO correction — but is that correct? Have we over-flattened our gathers in the NMO process and can't get to the true value of eta (or truly flat gathers)? Do our gathers even want to be flat? AVO effects can play tricks on us as illustrated by the phase change here:

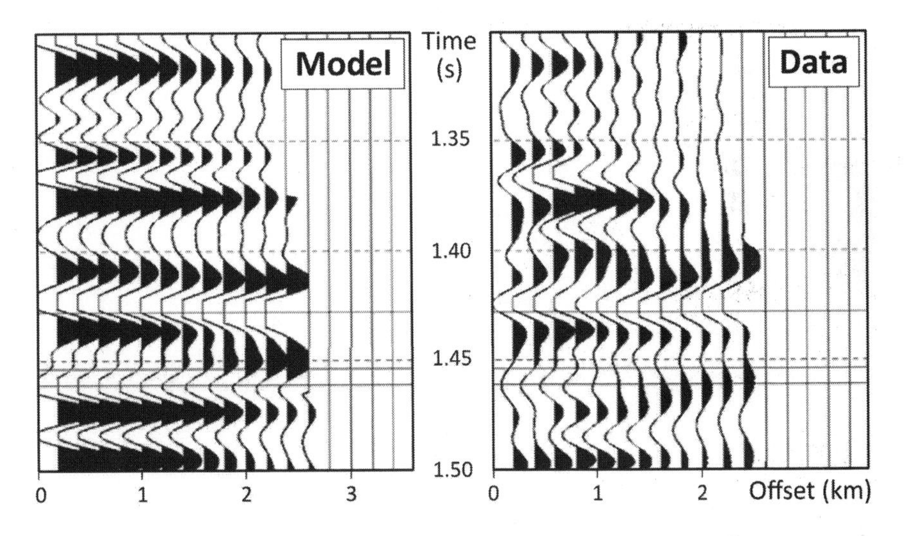

Signal processing, especially noise and multiple attenuation, have as many, if not more, questions involved with them. Even our friend deconvolution can damage an AVO response if it is applied trace-by-trace. Multiples are a persistent problem because they are often difficult to remove without damaging our signal, and can sometimes only become apparent way down the analysis or inversion route.

So a knowledge of the processing flow is a vital part of any interpretation exercise. It sets expectations as to what can be achieved. Ideally, the interpreter can get involved in the processing exercise and become familiar with the limitations in the seismic dataset, or some of the pitfalls. Failing that, a detailed processing report might be available which can help to answer some of these questions. If none of this direct or indirect knowledge is to be found, then perhaps reprocessing is the only option.

Learn to program

Matt Hall

Two years ago my eldest daughter came home from school and said she'd been programming robots. Programming robots. In kindergarten.

For the first time since I was four years old, I wished I was five.

Most people I meet and work with do not know how to make a computer do what they really want. Instead, they are at the mercy of the world's programmers and — worse — their IT departments. The accident of the operating system you run, the preferences of those that came before you, and the size of your budget should not determine the analyses and visualizations you can perform on your data. When you read a paper about some interesting new method, imagine being able to pick up a keyboard and just try it, right now… or at least in an hour or two. This is how programmers think: when it comes to computers at least, their world is full of possibility.

I am not suggesting that all scientists should become programmers, coding, testing, debugging, with no time left for science. But I am suggesting that all scientists should know how computer programs work, why they work, and how to tinker. Tinkering is an underrated skill. If you can tinker, you can play, you can model, you can prototype and, best of all, you can break things. Breaking things means mending, rebuilding, and creating. Yes: breaking things is creative.

But there's another advantage to learning to program a computer. Programming is a special kind of problem-solving and rewards thought and ingenuity with the satisfaction of immediate and tangible results. Getting it right, even just slightly, is profoundly elating. To get these rewards more often, you learn to break problems down, reducing them to soluble fragments. As you get into it, you appreciate the aesthetics of code creation: like equations, computer algorithms can be beautiful.

The good news for me and other non-programmers is that it's never been faster or simpler to give programming a try. There are some amazing tools to teach novices the concepts of algorithms and procedures; MIT's Scratch and App Inventor projects are leaders in that field. Some teaching tools, like the Lego MINDSTORMS robots my daughter uses, are even capable of building robust, semi-scientific applications.

Start on AWK now and you'll be done by lunchtime tomorrow.

Chances are good that you don't need to install anything to get started. If you have a Mac or a Linux machine then you already have access to scripting languages such as the shell, AWK, Perl, and Python — just fire up a terminal. On Windows you have Visual Basic. There's even a multi-language interpreter online at *codepad.org*. These languages are great places to start: you can solve simple problems with them very quickly and, once you've absorbed the basics, you'll use them every day. Start on AWK now and master it by lunchtime tomorrow.

Here are some tips:

- Don't do anything until you have a specific, not-too-hard problem to solve. If you can't think of anything, the awesome Project Euler has hundreds of problems to solve.
- Choose a high-level language like Python, Perl, or JavaScript; stay away from FORTRAN and C.
- If you already know or used to know MATLAB, try that — or the free alternative GNU Octave.
- Buy just one book, preferably a thick one with a friendly title.
- Don't do a course before you've tinkered on your own for a bit, but don't wait too long either.
- For free, online, world-class programming courses, check out *udacity.com.*
- Learn to really use Google: it's the surest and fastest way to find re-usable code and other help.
- Have fun brushing up on your math, especially trig, vectors, time series analysis, and inverse theory.
- Share what you build: give back and help others learn.

Acknowledgments

This essay first appeared as a blog post in September 2011, *ageo.co/HNxStU*

Leonardo was a geophysicist

José M Carcione

The science of geophysics studies the physics of our planet, considering the atmosphere, the hydrosphere, and the core, mantle, and crust of the earth. It is highly interdisciplinary since it involves geology, astronomy, meteorology, physics, engineering, and scientific computing. Today, it is impossible for a single researcher to deal with all these fields.

Before the scientific method was introduced, Leonardo da Vinci (1452–1519), one of the brightest minds of all time, excelled in every aspect of art, humanity, and science. Leonardo foresaw a number of geophysical phenomena. The list below is incomplete but illustrative of his discoveries.

Wave propagation, interference, and Huygens' principle (1678):

> *Everything in the cosmos is propagated by means of waves...*
>
> Manuscript H, 67r, Institut de France, Paris.

> *I say: if you throw two small stones at the same time on a sheet of motionless water at some distance from each other, you will observe that around the two percussions numerous separate circles are formed; these will meet as they increase in size and then penetrate and intersect one another, all the while maintaining as their respective centres the spots struck by the stones.*
>
> Manuscript A, 61r, Institut de France, Paris.

The Doppler effect (1842):

> *If a stone is flung into motionless water, its circles will be equidistant from their centre. But if the stream is moving, these circles will be elongated, egg-shaped, and will travel with their centre away from the spot where they were created.*
>
> Manuscript I, 87, Institut de France, Paris.

Newton's prism experiment (1666):

> *If you place a glass full of water on the windowsill so that the sun's rays will strike it from the other side, you will see the aforesaid colours formed in the impression made by the sun's rays...*
>
> Codex Leicester, 19149r, Royal Library, Windsor.

Explanation of the blue sky, before Tyndall's 1869 experiments and Rayleigh's theory of 1871:

> *I say that the blue which is seen in the atmosphere is not given its own colour, but is caused by the heated moisture having evaporated into the most minute and imperceptible particles*
>
> Codex Leicester, 4r Royal Library, Windsor.

The principle of the telescope, first constructed in the Netherlands in the early 17th century:

> *It is possible to find means by which the eye shall not see remote objects as much diminished as in natural perspective...*
>
> Manuscript E, 15v, Institut de France, Paris.

> *The further you place the eyeglass from the eye, the larger the objects appear in them*
>
> Manuscript A, 12v, Institut de France, Paris.

> *Construct glasses to see the Moon magnified.*
>
> Codex Atlanticus, 190r, a, Ambrosiana Library, Milan.

A statement anticipating Newton's third law of motion (1666):

> *As much pressure is exerted by the object against the air as by the air against the body.*
>
> Codex Atlanticus, 381, Ambrosiana Library, Milan.

The principle of least action, before Fermat in 1657 and Hamilton in 1834:

> *Every action in nature takes place in the shortest possible way.*
>
> Quaderni, IV, 16r.

The evolution of the earth and living creatures, preceding George Cuvier (1804) and Charles Lyell (1863), and plate tectonics, anticipating Wegener (1915):

> *That in the drifts, among one and another, there are still to be found the traces of the worms which crawled upon them when they were not yet dry. And all marine clays still contain shells, and the shells are petrified together with the clay.*
>
> *...Strata were covered over again from time to time, with mud of various thickness, or carried down to the sea by the rivers and floods of more or less extent; and thus these layers of mud became raised to such a height, that they came up from the bottom to the air. At the present time these bottoms are so high that they form hills or high mountains, and the rivers, which wear away the sides of these mountains, uncover the strata of these shells,...*
>
> Codex Leicester, Royal Library, Windsor.

Mind the quality gap

Pavlo Cholach

A present-day geophysicist faces a world of increasing pace, information overload, and a shrinking and aging workforce. Timely delivery of high-quality geophysical interpretation products (e.g. forward models, well-ties, time horizons, depth converted maps and volumes, well prognosis, pore pressure prediction, inversions for reservoir properties, geobodies) in these settings is paramount. A commonly circulated view is that a combination of improved geophysical tools, standardization, and increased efficiency provides the way to meet this challenge. Geophysical tools are delivered by a variety of software companies annually providing ever more sophisticated upgrades. Standardization is addressed through 'best practice' and 'approved' workflows — concepts that are engraved in the psyche of multinational exploration and production company employees. Achieving higher efficiencies appears to be the most elusive challenge. Efficiency requires a balance between the quality of interpretational products and the time it takes to deliver them.

Established paradigm stipulates that with more time available, the quality of interpretations would improve. Even though extra time provides an opportunity to deliver better products, the relationship between time and quality could be non-linear and complex, as shown here:

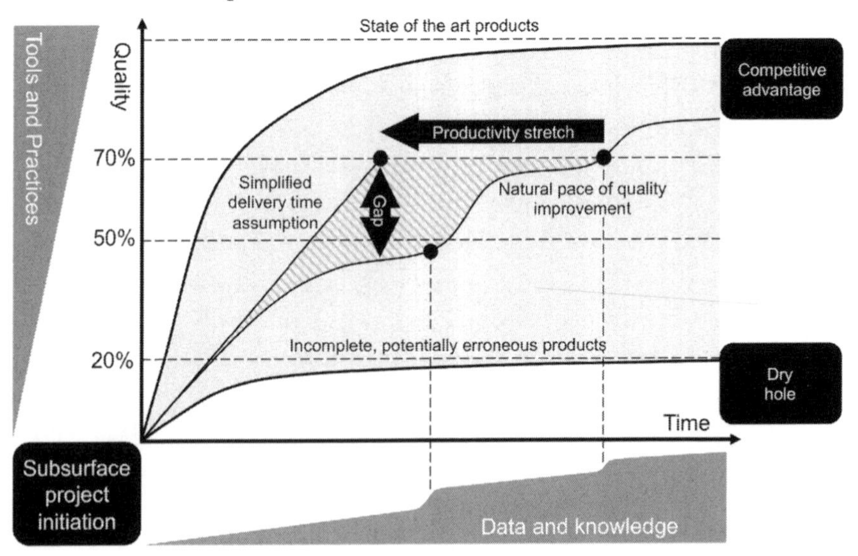

With shortened delivery time, there must be a deterioration in quality, but this quality gap is poorly understood and difficult to quantify.

By not fully recognizing how quality could be achieved in a timely manner, a geophysicist faces so-called efficiency stretch almost permanently. We may use phrases like 70 percent solution in an attempt to describe products delivered on artificially compressed (i.e. arbitrary) timelines and to justify potential quality issues. Furthermore, due to extremely tight timelines in our business, efficiency stretch is often a way to drive performance and achieve the 70 percent solution in the shortest time possible. In other words, a simplified linear assumption about the timing of subsurface products prevails.

It's inevitable that with an aging geophysical workforce and a shortage of skilled geophysicists this trend will only accelerate. The simplified assumption line leading to the 70 percent solution will simply steepen. It is critical for us to recognize what this means for the quality of interpretations. With shortened delivery time, there must be a deterioration in quality, but this quality gap is poorly understood and difficult to quantify. As shown in the chart, productivity stretch could lead to a significant drop in product quality — a 70 percent stretch solution might only yield 50 percent quality. This prospect should be alarming because poorer quality geophysical products eventually lead to dry holes, one of the biggest expenses in many conventional and some unconventional plays. Poorly performing wells are another outcome. After all, the geophysical contribution to exploration and production projects is to propose the best possible locations to drill, and geophysical products of high quality and reliability are essential in achieving excellent results.

Highly efficient tools and best practices, in combination with sufficient quality and quantity of data and knowledge of the play, will aid a geophysicist in the delivery of great products, and provide an exploration and production organization with a competitive advantage in today's ever more complex search for hydrocarbons. Chasing better efficiency in a productivity stretch is not a good strategy and just might open a quality gap.

My geophysical toolbox, circa 1973

Dave Mackidd

I pulled up in front of the old Esso building in Calgary, adjusted my tie, gulped, and walked in. After a bit of an orientation, I was introduced to my first boss. My training assignment would be the Canadian Arctic, where Esso was actively drilling onshore and offshore. Eager to make an impression as a worldly wise professional, I asked, 'Will I get to work on Prudhoe Bay?' My new boss exhaled, leaned back, and said, 'No. That's in Alaska.'

There was only one way to go from that auspicious start, and that was up.

New graduates will find 1973 technology hard to imagine. No cell phones, no computers, crude calculators. Seismic interpretation was done on big rolls of paper on long tables using coloured pencils and a large box of erasers. We each had a big draftsman's brush to remove eraser debris. Interpreters had stiff necks from leaning over the table to look sideways at the section: it was the only way to change the horizontal scale.

We had to be creative. I remember one geophysicist using a razor blade to cut his section into individual traces, then gluing it back together flattened on a reflector. Don't laugh! That method found a new Zama Reef pool in northern Alberta.

There's more than one way to flatten a section though, and there were some very innovative individuals at Esso in this analog age. One built a special camera for flattening. It was crude but effective: a geophysicist hand-picked a reflector on a film, which was run slowly through a large camera on a movable platform. The operator tracked the horizon, using a mechanical lever to keep it centered in the view-finder. *Voilà* — a flattened section!

Filtering was a special challenge. Ernie Shaw and Jack Hogg devised a laser imaging system that split the reflection of a seismic image through a fluid with a precise refractive index. It was then fed through a crude filter cut from a sheet of tin, which blanked out some of the spectral components, then back through a reversal of the set-up. The result: the first analog *f-k* filtered sections.

Once complete, interpretation times were carefully measured with a pair of dividers, then painstakingly transferred from paper sections to basemaps, one shotpoint at a time. Often, even the basemaps had to be hand-drawn, and a

I remember one geophysicist using a razor blade to cut his section into individual traces, then gluing it back together flattened on a reflector.

single map could take weeks to produce. Of course, all contouring was done by hand. Once a map was deemed final, it was sent to the drafting department to be reproduced.

I was told that I was lucky to be living in modern times — we had 24-channel systems shooting up to six fold! In the old days, the veterans reminisced, a geophysicist accompanied the acquisition crew and made maps from picking shot records. This seems crazy, but most of the oil in Alberta was found that way.

Despite everything, my first well was a giant discovery, Issugnak. A great way to start! Except that it's still sitting there with a net present value of zero, and that won't change until the Mackenzie Valley pipeline is built. It transpired that Wes Hatlelid had mapped it a few years before me, but Esso had lost it! We spent two years searching for the field before we discovered it again, right where we'd left it.

How did this happen? Remember, this was way before GPS; we barely had satellites. Near-shore surveying was done by triangulation from shore-based transponders. I have painful memories of hauling 12-volt batteries on my back up to those stations, all located on the highest hills we could find. It turned out that one of the transponders was giving spurious readings, so the survey for that summer's shooting was wrong. Take nothing for granted when you look at old data.

No more innovation at a snail's pace

Paul de Groot

In the mid 1970s I went to Delft University to study applied geosciences or mining engineering as it was called at the time. I was attracted by the sky-is-the-limit mentality of the oil men of the era. This was the heyday of exploration in the North Sea: oil was found almost daily and new fields were developed in turbulent seas using technology that had to be invented along the way. This was the vibrant, high-tech industry that I joined after graduation.

In the 30 years since, I have worked for a major oil company and a large R&D organization, and formed my own company. I have been fortunate to always work with the latest technology in my chosen field of seismic interpretation. This field has gone through a fantastic evolution. It is hard to imagine that when I started, a coloured pencil was the most important tool in the technical arsenal of a seismic interpreter. And look at us now — we immerse ourselves in a 3D model and steer a horizontal well through a thin layer of reservoir rock to extract the last drop of oil for an energy-hungry world. This is cool stuff that surely supports the image of a vibrant, high-tech industry. But are we indeed so vibrant and high-tech?

Personally, I don't think we are. We certainly have great technology, but it takes ages for innovations to become accepted and to be used widely. A former R&D director of the major oil company I worked for once told me that it takes 10 years from invention to production mode. Sadly, I believe he was right. And he's still right today.

It seems to me that when it comes to innovation, ours is a very conservative business. Only a handful of geoscientists and companies stick their neck out to try something new. Most prefer to wait for new technology to be proven time and again before they adopt it themselves. In my experience early adopters are more successful than followers, which makes me wonder: why are we not innovating at a much faster pace?

The following analysis is not complete and it has no scientific foundation. I merely highlight a few factors that may explain why innovation in the field of seismic interpretation is going at a snail's pace, and suggest what we can do to speed it up.

But are we indeed
so vibrant and high-tech?
Personally, I don't think we are.

- **Demography.** The G&G population is skewed with lots of old guys (like me). Most prefer to press the same buttons they have pressed for the last 20 years. Only few are willing to learn new tricks. *Be one of them!*
- **IT departments.** In large companies IT departments control rather than service the user community. Standards are important for IT people because standards make life easy for them. New technology does not fit into standards hence is blocked by them. In the rigorous drive towards standards there is no room for new technology that subsurface specialists need to find and produce our precious commodity. *Help your IT department see the big picture by showing how you will help meet standards with new technology.*
- **HSE consciousness.** The emphasis on HSE has made our industry a much safer place for people and the environment. I cannot agree more: HSE should be a top priority in everything we do. Still, I wonder whether we have changed the mentality of our work force somewhere along the line such that no one dares to take any risks at all. The risk-seeking spirit of the pioneers who built this industry has gradually been replaced by office workers who are not willing to stick out their necks and try something new. *Learn to recognize when the reward is greater than the risk.*

Old physics for new images

Evan Bianco

At its core, seismology is concerned with how objects move when forces act on them. Over 300 years ago, two gentlemen outlined everything we need to know: Robert Hooke, with his law describing elasticity, and Isaac Newton with his second law describing inertia. Anyone working with seismic data should try to develop an intuitive understanding of their ideas and the equations that manifest them.

For rocks, a rudimentary but useful analogy is to imagine a mass suspended by a spring. Hooke discovered that when the spring is stretched, stress is proportional to strain. In other words, the force vector **F** exerted by the spring is proportional to the magnitude of the displacement vector **u**. The proportionality constant k is called the stiffness coefficient, also known as the spring constant:

$$\mathbf{F} = -k\mathbf{u}$$

This is the simplest form of Hooke's law of elasticity. Importantly, it implies that the stiffness coefficient is the defining property of elastic materials.

Newton's second law says that a body of mass m, has a resistance to acceleration $\ddot{\mathbf{u}}$ (that is, the second derivative of displacement with respect to time) under an applied force **F**:

$$\mathbf{F} = m\ddot{\mathbf{u}}$$

If displaced from equilibrium, a mass attached to the end of a spring will feel two forces: a tensional force described by Hooke's law, and an inertial force from its motion, described by Newton's second law. The system of a mass and a single spring yields simple harmonic motion, characterized by acceleration being proportional to displacement but opposite in direction:

$$m\ddot{\mathbf{u}} = -k\mathbf{u}$$

Simple harmonic motion has many applications in physics, but doesn't quite fit the behaviour of rocks and seismic waves. A rock is bounded, like a mass held under the opposing tension of *two* springs. In this case, there are two tensional forces acting in the line along which the mass can oscillate. Writing

Rock properties dance upon the crests of travelling waves, and they dance to the tune of seismic rock physics.

out the forces in this system and doing a bit of calculus yields the well-known wave equation:

$$\ddot{\mathbf{u}} = \frac{k}{m}\nabla^2\mathbf{u}$$

The wave equation says the acceleration of the mass with respect to time is proportional to the acceleration of the mass with respect to space, a tricky concept described by the Laplacian ∇^2. The point is, the only properties that control the propagation of waves through time and through space are the elasticity of the springs and the inertia of the mass.

Some vector calculus can move our spring–mass–spring system to three dimensions and unpack k, m, and ∇^2 into more familiar earth properties:

$$\ddot{\mathbf{u}} = \frac{\mathbf{F}}{\rho} + \left[\frac{\lambda + 2\mu}{\rho}\right]\nabla(\nabla \cdot \mathbf{u}) - \left[\frac{\mu}{\rho}\right]\nabla \times (\nabla \times \mathbf{u})$$

Here, λ and μ are the Lamé parameters, representing Hooke's elasticity, and ρ is the density of the medium, representing Newton's inertia. You don't need to fully comprehend the vector calculus to see the link between wave mechanics, as described by the displacement terms, and rock properties. I have deliberately written this equation this way to group all the earth parameters in the square brackets. These terms are equal to the squares of P-wave velocity V_P and S-wave velocity V_S, which are therefore nothing but simple ratios of tensional (λ and μ) to inertial properties (ρ).

To sum up, the Lamé parameters and density are the coefficients that scale the P-wave and S-wave terms in the wave equation. When rock properties change throughout space, the travelling waveform reacts accordingly. We have a direct link between intrinsic properties and extrinsic dynamic behaviours. The implication is that propagating waves in the earth carry information about the medium's intrinsic parameters. Rock properties dance upon the crests of travelling waves, and they dance to the tune of seismic rock physics.

One cannot live on geophysics alone

Marian Hanna

Many geophysicists make the mistake of not broadening their understanding of geology, engineering, and economics. A geophysicist may not understand or be able to convince others of the true value of what geophysics has to offer without that broader perspective. But the reality is that if we integrate multi-disciplinary data deep into our work, we will have a thorough approach to our solutions.

So what does this mean to the geophysicist? At university, enroll in courses outside your core curriculum. Consider economics, business, geology, and engineering electives. A psychology course or two wouldn't hurt either. Enter the AAPG's Imperial Barrel Award or encourage your department or company to hold a similar workshop incorporating geology, geophysics, engineering, and business. This type of workshop is often best positioned for final-year or postgraduate students or any employee in an oil and gas company. The premise is to have one person of each discipline on a team. Each group is given a fictitious budget and with that money the teams compete for exploration licenses, acquire or process seismic, drill a well, and put the well on production while looking at the rate of return on the investment and any other financial drivers, including booking reserves and resources. This is all done in competition with the other teams and everyone receives real well data, seismic, and production data as they work through the steps. Sounds like real life in the oil and gas industry, doesn't it?

There may be many times in your career that you can look back and smile at such an assignment. Multi-disciplinary teamwork can bring some of the most frustrating and most exciting experiences. Why exciting? It can be groundbreaking and it is often the kind of effort required to conquer major hurdles in a company's exploration program. Why frustrating? Because you not only have to generate ideas, but you need to convince your teammates and management of your approach and how it integrates all your data. That is not always easy — your idea may be dismissed readily by those that don't quite see your vision. Hang in there! Remember that it is always easier to shoot an idea down than come up with something original. If your idea is truly a good one then don't give up. As the saying goes, a new idea is like a child. It's easier to conceive than to deliver.

A new idea is like a child.
It's easier to conceive
than to deliver.

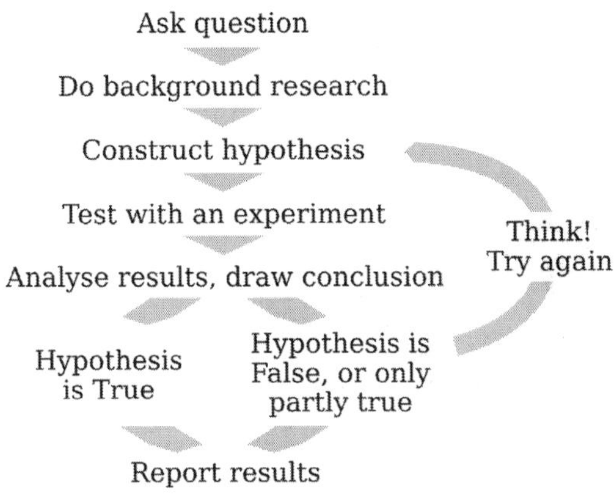

Ask question
Do background research
Construct hypothesis
Test with an experiment
Analyse results, draw conclusion
Hypothesis is True
Hypothesis is False, or only partly true
Think! Try again
Report results

If you can prove your point with scientific methods then you are on to something big. And you will know your idea is fantastic when others attach themselves to your work, or copy it. There is an old saying that acting on a good idea is better than just having a good idea. So start working hard on your next great idea!

Pick the right key surfaces

Mihaela Ryer

Seismic stratigraphy is an approach that allows the geoscientist interpreter to extract stratigraphic and depositional environment information and insight from seismic data. One of the critical steps in any rigorous seismic stratigraphic workflow is the definition of key surfaces, which form the basis and the framework for any further qualitative or quantitative analysis.

Key surfaces are seismic reflectors of regional extent that can be defined and mapped on the basis of seismic reflection terminations, internal reflection configuration, and rigorous integration with any available well data. Five principal types of reflection terminations are associated with key seismic stratigraphic surfaces: onlap, downlap, toplap, truncation, and concordant. The internal reflection configuration is the seismic proxy for sedimentary stratal patterns; recording and documenting it gives valuable insight into depositional processes and products. Commonly six types of reflection configurations are used for seismic key surface definition: parallel, subparallel, divergent, prograding, chaotic, and reflection-free.

Types of seismic key surfaces

Seismic stratigraphic key surfaces are of two types: sequence boundary and maximum flooding surface (MFS); several criteria are used to properly define them.

Sequence boundary — probably the most abused key seismic surface. Inexperienced seismic stratigraphic interpreters may call any strong and/or continuous reflector a sequence boundary. A correct definition is not only an academic issue; it has significant implications for depositional history and stratigraphic and depositional environment predictions. We use the following criteria to define a sequence boundary:

1. onlap of coastal deposits onto the boundary;
2. erosion and incision on the shelf;
3. toplap or apparent toplap below the surface;
4. a major basinward shift in sedimentary facies across it.

Genetically, a sequence boundary is the result of an abrupt relative sea-level lowering, resulting in subaerial exposure, erosion, and fluvial incision of the

A correct definition is not only an academic issue; it has significant implications for depositional history and stratigraphic and depositional environment predictions.

shelf. During this time, sediment generally bypasses the shelf area and is deposited on the slope or basin floor via gravity-flow processes.

Maximum flooding surface — a continuous draping event resulting from regionally extensive and significant water deepening. It is commonly recognized on the seismic because of downlap reflection terminations onto the surface. Maximum flooding surfaces are commonly overlain by a regionally extensive marine mudstone or shale and, as a result, abrupt changes in incremental overpressure are present across them. Genetically, maximum flooding is the result of a rapid relative sea-level rise resulting in the shoreline migrating to its maximum landward position.

Maximum flooding surfaces and sequence boundaries are the key bounding surfaces in a depositional sequence. They are used to define the systems tracts which, together with the depositional environment and sedimentary facies, are used to interpret the sedimentary evolution of a basin or area. If the sequence stratigraphic analysis is used in the context of a petroleum system evaluation, the proper key surface definition and interpretation plays a major role in the prediction of reservoir, seal, and source rock.

For lots more on seismic stratigraphic key surfaces, or seismic sequence stratigraphy in general, visit the SEPM Sequence Stratigraphy Web at *sepmstrata.org*.

Practise pair picking

Evan Bianco

Imagine that you are totally entrained in what you are doing: focused, dedicated, and productive. If you've lost track of time, you are probably feeling *flow* (sedimentologists might specify laminar, rather than turbulent, flow). It's an awesome experience for an individual, but imagine the potential when a team feels it. Because there are so many interruptions that can cause turbulence, you are going to need some tricks. Seismic interpreters can seek out flow by partnering up and practising pair picking.

Having a partner in the passenger seat is not only ideal for training, but it is an effective way to get real work done. In other industries, pairing up has become routine because it works. Software developers sometimes code in pairs, and airline pilots share control of an aircraft. In both cases, one person is at the controls, the other is monitoring, reviewing, and navigating. One person for tactical jobs, one for strategic surveillance.

Here are some reasons to try pair picking:

Solve problems efficiently. If you routinely have an affiliate, you will have someone to talk to when you run into a challenging problem. Sticky workarounds become less tenuous when you have a partner. You'll adopt more robust solutions and be less prone to one-off hacks you can't remember a week later.

Integrate smoothly. There's a time for handover, and there will be times when you must use other people's work to get your job done. 'No! Don't use *Evan_Top_Cretaceous_final*, use *EB1_K_temp_DELETE-ME*.' Pairing with the predecessors and successors of your interpretation will get you better-aligned.

Minimize interruptionitis. If you have to run to a meeting, or the phone rings, your partner can keep plugging away. When you return you will quickly rejoin. It is best to get into a visualization room, or some other distraction-free room with a large screen, so as to keep your attention and minimize the effect of interruptions.

Mutual accountability. Build allies based on science, technology, and critical thinking. Your team will have a clearer group understanding of your work, and you'll feel more connected around the office. Is knowledge hoarded and privileged or is it open and shared? If you pick in pairs, there is always someone who can vouch for your actions.

Having a partner in the passenger seat is not only ideal for training, but it is an effective way to get real work done.

Mentoring and training. By pair picking, newcomers quickly get to watch the flow of work, not just a schematic and idealized flow-chart. Instead of just an end product, they see the clicks, the indecision, the iteration, and the pace at which tasks unfold.

Practising pair picking is not just about sharing tasks, it is about channelling our natural social energies in the pursuit of excellence. It may not be practical all of the time, and it may make you feel vulnerable, but pairing up for seismic interpretation might bring more flow to your workflow.

Acknowledgments

This essay first appeared as a blog post in June 2011, *ageo.co/LWwo87*

Practise smart autotracking

Don Herron

Horizon autotracking is one of the most critical tasks that we do when interpreting seismic data on a workstation. All interpretation systems contain functionality for this, but each is different in the sense that the user has varying degrees of control over tracking parameters. Three important considerations for using horizon autotracking are:

1. The data to be autotracked must have a sufficiently good signal-to-noise ratio so that you can reasonably expect autotracking to produce reliable results. It is your responsibility to assess overall data quality and signal:noise on a horizon-by-horizon basis before using autotracking.
2. Once you've decided to use autotracking, you should invest time in testing the sensitivity of autotracking output to changes in user-specified tracking parameters. These parameters include, but are not limited to: the time interval of the correlation window; the 'search distance' or amount of shift up or down along the seismic trace that the correlation window can be moved in making a pick; and the goodness-of-fit required to propagate the pick to the next traces. As you might guess, autotracking is iterative in the sense that you usually converge on an optimum set of tracking parameters as you evaluate tracking results and adjust input values. Having said this, you should also be aware that any given set of tracking of parameters will probably vary both laterally and vertically throughout a project because of changes in data quality and geology. Most autotrackers have functionality to allow the parameters to vary iteratively.
3. Within an interpretation workflow you should include time for quality control of tracking output. Autotracking results are acceptable only if the tracking algorithm makes its pick in the same place that you would. For the sake of both accuracy and efficiency you probably should not autotrack a horizon which you spend more time correcting than you would have spent, in the limiting case, picking every line in your dataset. Within the context and business constraints of your interpretation project you must decide on the degree of tracking accuracy you need to achieve acceptable results. This decision is not always obvious or easily made at the beginning of an interpretation, and can change as your work progresses.

Any given set of tracking of parameters will probably vary both laterally and vertically throughout a project because of changes in data quality and geology.

In addition to these three considerations, you should remember that autotracking operates by correlating from trace to trace on the basis of similarity of reflection character. Accordingly, there are some geologic surfaces, for example the walls of incised channels, which by their nature do not lend themselves to autotracking and must be manually picked. Your judgment in when and how to autotrack horizons will develop as you gain experience in the art of seismic interpretation.

Further reading

Herron, D A (2000). Horizon autopicking: *The Leading Edge* **19**, 491–492.

Pre-stack is the way to go

Marc Sbar

With the typical questions we need to answer today there is really no choice but to use pre-stack data. Of course it is only one of the elements you need for a thorough interpretation, whether you are dealing with rank exploration or development. The following discussion will explain why.

To start with you must request the right data. Whether you are reprocessing your seismic or buying it off the shelf, insist that the gathers and angle stacks are provided with the normal stacked data. The gathers should be migrated and, depending on the data range and fold, three angle stacks are best. To obtain the best balance among the angle stacks ask for the same fold in each, particularly in the target zone. Check that your final stacks, including the angle stacks, are created from the gathers that you receive. Sometimes the gathers are output at a later time and processed separately and scaled differently.

Let's leave the seismic for now and focus on another key source of data for your interpretation — well data. If you are fortunate enough to have a few wells, they can provide significant constraints on the rock properties in the target zone. If there is an opportunity to influence the logs acquired in the wells insist on a full suite that includes compressional and shear transit time and density along with other logs. Consider the questions you have to answer and which logs may best describe the rock properties of the zone of interest. For example, you may wish to learn about the anisotropy (dipole shear) in the formation or map fractures (high resolution resistivity).

So let's assume you have a complete log suite. Of course your trusty petrophysicist should bless the logs. This is an important data quality step and should be done before analysis. The next order of business is to tie the well or wells to the seismic. If you have a poor tie, then either or both the well and seismic may need more work. With a good tie in one or more wells the phase of the seismic can be established, so that interpretation can proceed on a correctly displayed dataset. Most people use zero phase data as their base volume.

The initial well tie is done in 1D. The next step is to match the seismic gathers with the model gathers from the well. In this step the amplitude scaling with offset or angle can be tested and corrected, if needed. Unless your processor

Whether you are reprocessing your seismic or buying it off the shelf, insist that the gathers and angle stacks are provided with the normal stacked data.

used well data in offset scaling, it will most likely have to be adjusted. This is critical if one is to attempt to extract rock property information from the seismic gathers. Don't forget to scale the angle stacks at the same time.

Now you are ready for some analysis. Forward modelling of gathers using the well data can demonstrate whether the property variation you seek can be identified. Some of the key rock properties that can be detected are variations in porosity, cementation, pore fluid, and lithology. They are not all independent or unique, so it is important to apply other geologic constraints to make the problem tractable.

Your software should enable you to calculate the amplitude of zero offset (intercept) and the change in amplitude with offset (gradient) on a target reflector. If there are measurable changes in intercept and gradient that can be related to something of interest, then create an AVO class volume from the angle stacks and use it to scan the data for the character change you desire. Use a program that links the class volume to the gathers, then inspect the gathers to verify the anomaly. Does it make geologic sense if you map it? Are there other ways to explain this observation? This is where experience in many different geologic settings is valuable.

Once the data quality has been verified and you have done some of the quick tests above, it makes sense to create pre-stack inversion volumes to obtain a better measure of uncertainty. Rock property volumes can then be computed based on relationships from the well data and rock property databases. These can more clearly define the lead, prospect, or area for development and help you determine the volume and assess the risk of drilling. You'll wonder why you ever drilled a well without it.

Prove it

Matt Hall

How many times have you heard these?

- The signal:noise is lower (or higher/improved/reduced)
- It's too thin to see (interpret/detect) on seismic
- You can't shoot seismic in the summer (winter/snow/wind)
- More fold (bandwidth/signal:noise/data) is too expensive
- That won't (will/can/can't) work
- It looks geological (ungeological/right/wrong)

I say these sorts of things all the time. We all do. We feel like we're bringing our judgment to bear, we're exercising our professional insight and experience. It's part of the specialist advisor role, which many of us play, at least from time to time. Sometimes, when time is short or consequences are slight, this is good enough and we can all move on to more important things.

Often though, we do have some time, or the consequences are substantial, and we need a more considered approach. In those situations, at least for most of us, it is not enough to trust our intuition. Our intuition is not a convincing enough reason for a decision. Our intuition is unreliable (Hall 2010).

Science is reliable. So challenge your intuition with a simple task: prove it. The bed is sub-resolution? Prove it. More fold costs too much? Prove it. This attribute is better than that? Prove it.

First, gather the evidence. Find data, draw pictures, make spreadsheets, talk to people. What were the acquisition parameters? What actually happened? Who was there? Where are the reports? Very often, this exercise turns up something that nobody knew, or that everyone had forgotten. You may even find new data.

Next, read up. Unless you're working on the most conventional play in the oldest basin, there has almost certainly been recent work on the matter. Check sources outside your usual scope — don't forget sources like the Society of Petroleum Engineers (*spe.org*) and the Society of Petrophysicists and Well Log Analysts (*spwla.org*), for example. Talk to people, especially people outside your organization: what do other companies do?

It is not enough to trust our intuition. Our intuition is not a convincing enough reason for a decision. Our intuition is unreliable.

Then, model and test. If you're measuring signal:noise, seismic resolution, or the critical angle, you're in luck: there are well-known methods and equations estimating those things. If you want to shoot cheaper data, or operate out of season, or convince your chief that vibroseis is better than dynamite, you'll have to get creative. But only experiments and models — spreadsheets, computer programs, or even just mind-maps — can help you explore the solution space and really understand the problem. You need to understand why shots cost more than receivers (if they do), why you can mobilize in June but not July, and why those air-guns in the swamp were a waste of time and money. Modelling and testing take time, but the time is an investment. Most projects are multi-year, sometimes multi-decade, initiatives. The effort you spend may change how you operate for many years. It's almost always worth doing. If your boss disagrees, do it anyway. You will amaze everyone later.

Finally, document everything. Better yet, do this as you go. But do wrap up and summarize. You aren't just doing this for the geoscientist picking up your project in five years, you're also doing it for your future self. Since most interpretation tools don't have built-in documentation capabilities, you'll have to find your own tools — look for ones that let you add links and comments for a rich report you can easily share. Wikis are perfect.

At each step, either find or pretend to be the most skeptical person in the world. Ask the tough questions. Doubt everything. Then prove it.

References

Hall, M (2010). The rational geoscientist. *The Leading Edge* **29** (5), 596 ff, DOI 10.1190/1.3422460

Publish or perish, industrial style

Sven Treitel

The old saw 'publish or perish' is often derogatively used to account for the flood of publications coming from certain members of the academic community. A different, and less humorous, interpretation of this term applies, I believe, to some of those doing industrial research in private corporations.

While academics are under ongoing pressure to publish to obtain promotions and research grants, industrial scientists often face the opposite problem: they are discouraged from publication by management fearing that release of significant technical know-how must invariably benefit the competition. This can happen when a manager, not sufficiently familiar with the subject of a paper requested for release, finds that the simplest way out is to say no. It is true that such restrictions are sometimes justified, but my experience over several decades in industrial R&D suggests that these concerns are usually unfounded. My own professional experience has been with a major oil company, but I venture to guess that the observations I make here are hardly unique to this industry.

What happens to a scientist working in this kind of an industrial environment? As the years roll by, he writes technical reports, which are read by a few of his coworkers, but the research never faces the scrutiny of peers on the outside. A successful scientist needs to interact with his professional colleagues through the vehicle of written as well as oral publication: those who do not do this tend to become professionally ossified over time. Of course the employer loses as well: an unmotivated and insular R&D staff is unlikely, even unable, to come up with cutting edge results.

There is an additional and equally nefarious consequence of a restrictive industrial publication policy: a scientist's worth in the job market is in large measure his publication record. Layoffs in industry have become an increasingly popular means to cut costs under the unrelenting pressure from investors. R&D often seems to be an early item to go on the block, and now the unknown, terminated industrial research scientist is left to fend for himself. He or she must compete with those better-known in their field by virtue of their publication record, and thus faces an uncertain and increasingly grim professional future. Clearly the best way for an industrial scientist to avoid falling into this trap is

Make certain that the prospective employer's practices include a reasonably open publication policy.

to make certain that the prospective employer's practices include a reasonably open publication policy.

From the employer's viewpoint, a reasonable publication policy makes even more sense: a company staffed by aggressive and creative scientists continuously interacting with their peers outside their own organization is much more likely to be successful over time than one which is obsessively secretive. A scientist remaining in such a restrictive environment is bound to perish professionally over time and lose marketability outside the company. As Matt Hall put it so aptly to me when I proposed this essay to him: 'It's ironic that preparing yourself to be laid off would probably lead to you not being laid off!'

Recognize conceptual uncertainty and bias

Clare Bond

How many different interpretations could be made of the same seismic image? This is a question that few interpreters consider when faced with a seismic image to interpret. That's because we try and match the image to something familiar and hone in on those qualities of the image that are most similar to those we expect. When we determine the concept that best matches the image, in our own eyes, we find it very difficult to move away from this initial concept and reflect on alternatives. Although not commonly considered in seismic image analysis the conceptual uncertainty associated with model choice during interpretation of seismic images, in combination with human biases, has the potential to severely influence the final subsurface model.

Consider the data used to produce a seismic image. It has a limited resolution, like any remotely sensed data, and an inherent uncertainty. In fact each element in the process from seismic acquisition to processing has associated decisions that narrow the potential conceptual interpretations that are most likely for the dataset. For example, the type of acquisition chosen may be influenced by, amongst other things, the terrain, economics, and practicalities, but also by the depth and structural style of the target. The processing also leaves its imprint, with the chosen parameters having a direct impact on the features highlighted in the final seismic image. These non-economic and practical decisions are based on prior knowledge that itself has differing levels of certainty, or uncertainty.

On top of this uncertainty in the data is the influence of human bias at the interpretation stage. Humans are prone to using heuristics (rules of thumb) that allow the processing of large volumes of complicated data quickly — cunning short-cuts. They are very powerful, but not always correct. For example, imagine listening to a telephone conversation at the far end of an open-plan office; from the snippets heard it is possible to use these to construct the full conversation. Or at least the conversation expected based on the elements heard and any other prior knowledge you can bring into play (such as person *X* has just split-up with person *Y*). When we interpret we use all our prior knowledge consciously or sub-consciously to help us: the geographical location, first-hand knowledge of the tectonic or sedimentary regime, and second-hand knowledge from papers

Sometimes heuristics can set us on completely the wrong track, and... it is hard to turn around.

and books, or what others have said. This prior knowledge helps us to fill in the 'holes' in the data presented to us, it helps filter all the possible concepts to those that are most likely. We use it to focus on the elements of the data that are most important to confirm the concepts in our heads.

But sometimes heuristics can set us on completely the wrong track, and once heading in the wrong direction it is hard to turn around. Some of the human biases that can affect interpretational outcome and our ability to use heuristics effectively are:

- **Availability bias.** The decision, model, or interpretation that is most readily brought to mind.
- **Confirmation bias.** To seek out opinions and facts that support one's own beliefs or hypotheses.
- **Anchoring bias.** Failure to adjust from experts' beliefs, dominant approaches, or initial ideas.
- **Optimistic bias.** 'It won't happen to me' mentality or 'there is definitely oil in this prospect!'.

Failure to consider the other possible interpretations or concepts, the full breadth of the conceptual uncertainty challenge, could mean a dry well or a missed opportunity.

Further reading

Bond, C E, A D Gibbs, Z K Shipton, and S Jones (2007). What do you think this is? 'Conceptual Uncertainty' in geoscience interpretation. *GSA Today* **17**, 4–10, DOI 10.1130/GSAT01710A.1.

Krueger, J I, and D C Funder (2004). Towards a balanced social psychology: Causes, consequences and cures for the problem-seeking behaviour and cognition. *Behavioural and Brain Sciences* **27**, 313–327.

Rankey, E C, and J C Mitchell (2003). That's why it's called interpretation: Impact of horizon uncertainty on seismic attribute analysis. *The Leading Edge* **22**, 820–828, DOI 10.1190/1.1614152.

Tversky A, and D Kahneman (1974). Judgement under uncertainty: heuristics and biases. *Science* **185**, 1124–31.

Remember the bootstrap

Tooney Fink

One of the more challenging (and satisfying) experiences is using your seismic data to create an accurate prognosis for your proposed well. Typically this exercise involves a top-down analysis, starting with near surface geologic markers and working one's way deeper through the stratigraphic column. A few of the nearest wells are used, establishing ranges for formation tops as well as velocities for converting relevant seismic picks to depth.

How about building a prognosis from the bottom up? I imagine starting with a deeper marker, whose depth might be more predictable than first thought, or at least more predictable than a complex overburden above the target zone. How can a deeper marker be more predictable? In an intercratonic basin like the Western Canadian Sedimentary Basin the Palaeozoic, for instance, dips steadily and predictably from the foreland into the foredeep of the Rocky Mountain thrust belt. Palaeozoic depth structure maps, based on regional well control, can provide a relatively accurate datum below your target zone. Seismic data can be used to create an isochron from this deeper datum up to one's target, and converting this isochron to an isopach yields a depth to one's target. This approach has been used by Foothills explorationists for decades, especially when complex structuring in the overburden challenged a more conventional top-down prognosis. Not only do you end up with a prognosis, but you've also created a regional structure map on a deep marker, also characterizing the regional strike at this stratigraphic level.

When working with deeper datums it is true that one is dealing with generally higher velocity sections, where two-way seismic time uncertainty means greater isopach uncertainty. In compensation, these higher velocity sections tend to be more vertically and areally consistent, and have a more consistent lithology, and are thus more predictable.

Another Western Canadian Sedimentary Basin example that can use a bottom-up prognosis is the Mississippian subcrop play family. The deeper datum often used here is the Mississippian–Devonian boundary of the Exshaw, a good seismic marker. Once again, the isochron from the Exshaw up to one's prospect top can yield an isopach, which can accurately predict how much Banff, or Pekisko, or Shunda, or Elkton Formation (or all of these) is present at your proposed

How about building a prognosis from the bottom up? I imagine starting with a deeper marker, whose depth might be more predictable than first thought...

location. Just as important as the depth prediction is the determination of the stratigraphic interval left at the Mississippian subcrop, some of these intervals being more prone to porosity development than others.

What other seismic phenomena might be visible and predictable in depth or stratigraphic level?

- **Detachment surfaces** in thrust terranes are often stratigraphically consistent, occurring in the more ductile intervals of the section (e.g. anhydrites in the Wabamun, ductile shales such as the Exshaw or Nordegg). Such detachments may be identified on seismic, confirming a stratigraphic level, hence aiding in the prognosis.
- **Synclines**, in thrusted terrains, are not only 'pin' points for palinspastic restoration, but are also imaged preferentially on seismic, and their depth (at their base) may be prognosed from the sum of all the overlying section, based on both well control and surface outcrop measurements (a powerful combination).
- In extensional environments, growth-style **sole faults** will often sole out in a common stratigraphic level.
- **Salt welds** may also occur at stratigraphically consistent levels, at the base of the original salt layer or the base of the remobilized salt pillow.
- **Metamorphic or volcanic zones** are usually good seismic markers, and their depth may be predicted using non-seismic data such as gravity, magnetics, or magnetotellurics.

Yes, many of these deep markers may have larger uncertainty in their depth predictions, as a result of very sparse well control at these depths, but it is always instructional to examine your prognosis from the bottom up, as a complement to your more conventional top-down prediction.

Resolution on maps and sections

Rob Simm

Every interpreter knows that the bandlimited nature of seismic data limits its resolving power. We have all seen the picture of a rock face with a seismic wavelet superimposed, giving a salutary reminder of just how difficult the task of connecting seismic with geology is going to be. But what is resolution? Much of the time we think of 'vertical resolution', the notion that there is a minimum thickness of a geological unit for which the top and base can be uniquely identified. Of course, Widess (1973) is required reading for every young interpreter and the quarter-wavelength criterion for tuning thickness should be seared into their brains. Sheriff (1977) showed that there is also a spatial dimension to the problem, that the seismic reflection from a boundary is the result of constructive interference over an area of the wavefront (the Fresnel zone) defined by the quarter wavelength. This was a serious issue when we generated prospects using unmigrated 2D data, requiring map migration by hand!

So it is all down to wavelength; no wonder then that the favourite mantra of the processing gurus at Britoil in the early 1980s was 'resolution is bandwidth'. 'Must remember that' I thought, but what is bandwidth? 'Octaves,' they muttered and the seismic inversion experts nodded sagely, knowing that more octaves on the low end means better inversions. But, like all fundamental issues, there is more than one way of looking at it. Octaves, yes, but also (and critically) wavelet shape. Koefed (1981) taught us that. It has also become clear from an understanding of AVO that resolution can also depend on stacking angle. Fascinating…

3D acquisition and migration was a massive game changer, apparently consigning the Fresnel zone to history. The dramatic improvements in imaging can be considered an improvement in resolving power, but the main benefit of 3D is the increased spatial context of one reflection to another. All the advantages of 3D analytical techniques have accrued from the ability to present the data in map form. Increasingly, it has become clear that the very definition of 'resolution' needs to be broadened to include the detection of features on maps which can be assigned a geological significance.

A favourite illustration is an early time lapse model published by Archer et al. (1993). A simple synthetic 3D model was constructed to model the effect of

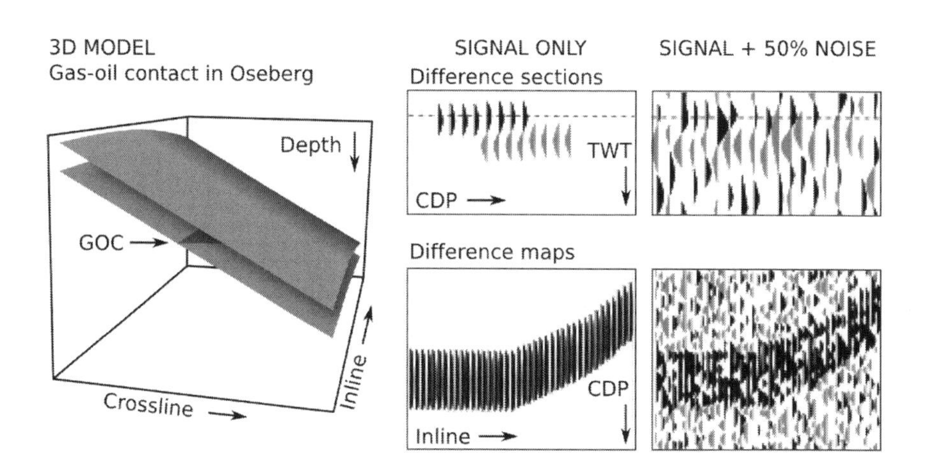

movement in the gas–oil contact (GOC) in the Oseberg reservoir. To make the results realistic, noise was added to the model such that the signal:noise ratio is 1.5 (pretty severe but never mind). The 'signal only' difference section shows that the expectation is a simple tuning response, but on the noisy section it is difficult to appreciate quite where the signal is. However, a time slice through the difference cube (position shown by the dashed line in the section) shows that even in the presence of noise the map lineaments associated with the contact rise can be interpreted (almost exactly). This is a great illustration of the increased dynamic range of maps over sections, and it is a good lesson to learn that interpretation is an iterative process between vertical sections and maps. Don't treat 3D as glorified 2D and spend months looking at vertical sections before making maps of amplitude and other attributes.

References

Archer, S, G King, R Seymour, and R Uden (1993). Seismic reservoir monitoring — the potential. *First Break* **11**, 391–397.

Koefed, O (1981). Aspects of vertical seismic resolution. *Geophysical Prospecting* **29**, 1–30.

Sheriff, R (1977). Limitations on resolution of seismic reflections and geological details derivable from them. In: Payton, C ed., *Seismic stratigraphy — Applications to hydrocarbon explorations*. AAPG Memoir **26**, 3–14.

Widess, M (1973). How thin is a thin bed? *Geophysics* **38**, 1176–1180.

See the big picture

Brian Russell

In order to do a good job as a geophysicist it is important to be on top of a lot of important detail (see *Sweat the small stuff*). However, at the other end of the spectrum it is equally important to see the big picture. Since we live in an era of super-specialization this is increasingly hard to do. An integrated project in any area of geophysics involves data acquisition, modelling, analysis, and interpretation, and in the mid-70s every geophysicist was expected to be on top of all of these aspects. But, in the early 21st century, no one person can be a specialist in even a sub-set of these different areas. So how can you be expected to juggle the need to understand the details with the importance of seeing the big picture? I think that the answer to this lies in two directions: collaboration and self-education. Let me discuss these two different strategies.

Collaboration

As I mention in *Sweat the small stuff*, collaboration is important when you integrate your expertise with the expertise of others. In that approach, you simply rely on others for their detailed knowledge. However, in doing this, take the time to ask them about the important ideas and literature in their field. For example, although you may be familiar with the way a geophysicist looks at anisotropy, why not find out how a reservoir engineer looks at the problem? It will probably be quite different. As another example, you may be familiar with the seismic response of shale overlaying a carbonate reef, but what are the important parameters that a geologist is interested in? Again, they will be quite different to your ideas. While discussing the project, try to put yourself in the other specialists' shoes, and see what questions they might have about the way you are approaching the problem. Take all this information and try to expand your horizons. In other words, try to see the big picture.

Self-education

The next step in improving your knowledge of fields outside of your core area of expertise is to start educating yourself about these fields. This can be done in many ways. You can attend in-company seminars. You can sign up for industry short courses, assuming that you have an enlightened management that considers continuing education as part of your work experience.

We have never before been in an era of such super-specialization. However, we have also never been in an era where information is so readily available.

Continuing education courses are ever more accessible online. Or, and this is my preferred route, you can search out and read the key literature, both in industry journals and textbooks. I am a dinosaur who likes the feel of paper but, more and more, this material is available in electronic form, which is both cheaper and more portable. This will mean that the next time you are on a long flight you can turn on your preferred electronic tablet and educate yourself about some area of the business that is not in your field.

Life-long learning

We have never before been in an era of such super-specialization. However, we have also never before been in an era where information is so readily available via the internet. So at the same time that you are learning about new developments in your own area of specialty, embark on a life-long learning adventure to keep abreast of the big picture.

Seek out the biostrat

Alex Cullum &
Linn Margareth Johansen

The world around us has not always been as we see it today. For example, it is estimated that over 99 percent of all known species are extinct (Newman and Palmer 2003). Humans live a relatively short life, so it is hard for us to conceptualize the processes that are at work on a geological timescale. Tectonics, volcanism, erosion, redeposition, meteor impacts, cycles of orbital fluctuation, and changes in radiation from the sun all affect the atmosphere, oceans, and other global systems at a range of timescales. Life on the planet has always had its form, behaviour, abundance, and distribution delineated by these global and extraterrestrial forces. Together with evolution these forces drive a turnover in the species living on the planet through inception and extinction.

Microfossil data from samples taken through sedimentary successions clearly show the evolution, extinction, and abundance changes of species through geological time. The evolution and extinction of a species does not provide an absolute age in millions of years, but occurs at the same point in time in datasets from other wells or outcrops. Fossil events and zones created from these can be used to subdivide and correlate stratigraphy between datasets. Models of geological time and stratigraphy (Gradstein et al. 2004; Ogg et al. 2008) allow biostratigraphic events and subdivisions to be aligned with absolute time.

While working with seismic cross sections seek out biostratigraphic information and plot it at well locations to aid interpretation. The stratigraphic units, ties, picks, and attributes identified on seismic may be relatively isochronous within the study area, in which case the biostratigraphy can be used as a quality check of the model being built. If the biostratigraphy shows that units are crossing time, then the extent to which this fits existing models and sequence stratigraphic simulations should be considered. Significant age differences where a seismic tie passes through a pair of wells might indicate that mis-ties have occurred. In such cases faults or unconformities might lie between the well locations, not yet resolved or observed in the seismic image.

Biostratigraphic analysis can be used to provide an independent interpretation of the depositional environment, water depth, and relative distance from land. This can be a useful input to those predicting rock types (for example source

Significant age differences where a seismic tie passes through a pair of wells might indicate that mis-ties have occurred.

rocks) from seismic data. Significant shifts in environment over short distances might indicate mis-ties or exceptional depositional settings.

With enough data, a log of geological age can be plotted against depth at well locations and overlain on seismic sections. Jumps in this log represent unconformities or faults which might otherwise not be apparent. When depth is on the vertical axis, the slope of this graph is proportional to sedimentation rate, low gradients on the plot representing low rates of sedimentation.

Biostratigraphy can deliver important insights before, during, and after drilling but like seismic interpretation, larger programs of analysis can take time. We suggest you instigate these projects as early as possible to ensure that results are available when you need them.

References

Gradstein, F M, J G Ogg, and A G Smith, editors (2004). *A Geologic Time Scale 2004*. Cambridge University Press.

Newman, M E J and R G Palmer (2003). *Modelling Extinction*. Oxford University Press.

Ogg, J G, G Ogg, and F M Gradstein (2008). *The Concise Geologic Time Scale*. Cambridge University Press.

Simplify everything

John Logel

The ability to simplify means to eliminate
the unnecessary so that the necessary may speak.
Hans Hofmann

In today's world of immediate and abundant information, it is easy to forget that geophysics is a specialized and complicated discipline that requires years of learning and understanding. It combines the sciences of mathematics and physics with the art of geology. This unique combination of left and right brain functions, along with the enthusiasm common in our field, can lead to very detailed, elaborate, sophisticated, and sometimes contorted explanations. We too often feel it is important to explain every detail and every equation, right from the fundamentals of our science. But sadly we can be misunderstood or, worse, not heard. By failing to pass on some of the elegance and beauty of it, we fail our science.

In our earliest math classes we learn to simplify. Fractions, algebraic equations, multiple equations, unknowns, and complicated word problems about trains and other physical features become more understandable and usable as we simplify them. Even matrix complexity can usually be reduced to very simple expressions.

One of the first lessons I learned in the oil business was from a supervisor named Tim Williams. After I tried to explain the then-young concepts of AVO, he told me, 'A lot of this technology isn't worth a damn, if you can't explain it simply.' I set out at that point to force myself to simplify.

Simplifying is not as easy as it sounds. You have to become quite involved in the science and mathematics of the process. You must fully comprehend and know the strengths, weaknesses, and assumptions of the process or theory, and then you have to break it into individual elements that are explainable through everyday occurrences or objects.

One way to simplify an idea is with an analogy. Some of my favourite geoscience simplifications are:

- Slinky used to show P-wave and S-wave propagation.

Simplifying is not as easy as it sounds. You have to become quite involved in the science and mathematics of the process.

- Marshmallows and sugar cubes to explain Young's modulus and Poisson's ratio for ductile and brittle rocks.
- Making coffee to explain shale gas diffusion (credit is due to my colleague Basim Faraj for that one).
- Clear squirt gun filled with sand and water to demonstrate overpressure.
- Loud neighbours to explain acoustic attenuation.

When you simplify any theory, concept, or geologic process, you make your ideas easier to understand, more engaging, and more memorable, too.

Simplifying does not just apply to science or to math but relates to everything we do. Making our lives simpler has the same effect. Things become easier to understand and more believable. People relate to you better and more often turn to you for help, advice, and guidance. Some of the best leaders demonstrate this ability. The best way we can represent our science and ourselves is to simplify everything.

As you simplify your life, the laws of the universe will be simpler; solitude will not be solitude, poverty will not be poverty, nor weakness weakness.
Henry David Thoreau

Sweat the small stuff

Brian Russell

Whoever said 'don't sweat the small stuff' had it wrong, and obviously wasn't in the geophysical business. (My wife once bought me a book by that name, and although I appreciate its advice for everyday life, I don't accept it in the workplace.) While it is important to see the big picture (see my essay *See the big picture*), it is equally important to worry about the small details that go into making that big picture. Our profession is made up of specialists and sub-specialists, and each one has a role to play in making sure that the final exploration decision is based on the fundamentals.

Typically, the integrated project involves the following steps: data acquisition, modelling, analysis, and interpretation. When I joined Chevron in the mid-70s, every exploration geophysicist was expected to be on top of all of these aspects of the project. But times have changed, and today no one person can be a specialist in all of these fields. So it is important to make sure that your contribution at the detailed level will fit with all of the other pieces to make an integrated whole.

An example

Suppose that your specialty is seismic data processing. An important new consideration for seismic processing is the effect of anisotropy. There are different types of anisotropy and different ways of analysing its effects, for example: in tomography, imaging, and velocity analysis. At your disposal you will have several seismic processing modules, based on some fundamental algorithms. You probably did not develop the algorithms or write the code (this is for other specialists) but this does not mean you don't have to understand how to use the modules effectively and how to optimize their parameters. This can seem like an overwhelming task. Unlike the situation 35 years ago, when a geophysicist only had to understand the basics of statics, the NMO equation, velocity spectra, post-stack time migration, and a few other fundamentals, the geophysicist of the 21st century has an amazingly complex set of concepts to master. So how would you approach this particular problem?

A suggested strategy

I would first recommend that you get as good a grasp of the fundamentals as

At the end of the process, you should be able to answer a number of questions, such as: 'why did you choose those parameters?'

you possibly can, either through reading the literature, attending a short course, or talking to in-company experts. Since you cannot be expected to become an expert on all the details, find out what details matter. For example, in the case of anisotropy, you will probably need to understand the Thomsen parameters, so make sure you have read and understood the original 1986 paper by Leon Thomsen. Next, find out how these details pertain to your particular task. This will involve finding the optimum way of extracting the correct parameters and applying them to your data. Finally, look at the results and make sure they make sense. This will be a collaborative task, which will probably involve discussing the result with both your colleagues and the client.

In summary, when I say 'sweat the details' I don't mean all the details, only the important ones. At the end of the process, you should be able to answer a number of questions, such as: 'Why did you choose those parameters?', 'How did it impact the final result?', and 'What should we do next?' Without understanding the important details, you will never be able to give good answers to these questions.

References

Thomsen, L (1986). Weak elastic anisotropy. *Geophysics* **51** (10), 1954–66. DOI 10.1190/1.1442051.

The evolution of workstation interpretation

Dave Mackidd

In 1977, I went to Houston to visit Exxon Production Research at the annual presentation of their research efforts. To a young geophysicist, this was indeed a plum assignment, but I had to present what I had seen to an amphitheatre of seasoned Esso veterans when I came back. Daunting as this was, I talked for about half an hour, then I summed it up by saying that this was not the most interesting thing that I had seen. During a break, I came upon a door with a window through which a light was shining. Upon peering in, I saw a velocity analyst picking gathers on a computer screen. Once he had picked one, he could press a button and the gather was redrawn flattened. The entire set-up had to be hard-wired through a thick cable to a mainframe computer less than 20 feet away. 'Some day', I said, 'we will put our coloured pencils and our erasers away and pick our seismic data on computer screens.' The hall erupted into laughter. That was the most outrageous thing they had ever heard!

My wild-eyed fantasy did not happen for some time and by that time I had moved to Canadian Hunter. The two main drivers for the emergence of the workstation were the advent of 3D seismic surveys and personal computers in the early 1980s. When 3D seismic arrived, we didn't know what to do with it. There was so much data! The processors would send us booklets of inlines and crosslines, and we would pick every tenth one in both directions (again with coloured pencils) and transfer that to a map. The rest (90 percent of the data) was ignored. Into the gap sprang Landmark Graphics, with the first commercial computer workstation for seismic interpretation, on a Sun Microsystems workstation running UNIX.

These machines were quite expensive, so only the larger companies could afford them. They were shared assets, so we had to book time on them. But they could do amazing things. We could choose the data we wanted to pick from a basemap on the screen. They had autopickers which could extend your picks on good reflectors to every line and trace in the volume. Your picks appeared on the screen in colour, and you could map the results. For the first time, geophysicists could make amplitude maps and infer lithology changes from those. This revolutionized interpretation, but the era of personal workstations was still a ways off.

'Some day,' I said, 'we will put our coloured pencils and our erasers away and pick our seismic data on computer screens.' The hall erupted into laughter.

However, I promptly marched in to my boss, the brilliant head of Canadian Hunter, John Masters, and asked if the company could buy one for me. John passed away in 2011 and was eulogized for his uncanny ability to see trends coming before anyone else. But this was perhaps the one time he missed the boat. I was told that they were too expensive and that he liked the way I interpreted data just fine as it was. So I decided to build my own workstation. Two wonderful guys helped me. One was George Palmer of Veritas Software, who wrote the interpretation code in an application he called Vista. The other was an electronics genius and alpha geek named Paul Geis. He designed the graphics card that could display seismic data. Together, we built a workstation on a 386 PC, circa 1984. It was state of the art with a 9-track tape reader for loading SEGY data, one of the first optical drives for storing it, and a Bernoulli Box with removable 20MB platters for storing horizons.

It wasn't perfect. It did not have the ability to work from a basemap — data was picked form a list. And it couldn't map data, instead picks were exported to a Perkin–Elmer minicomputer for contouring. However, I picked over 500 lines with it and Hunter made several oil discoveries in Saskatchewan as a result, including the prolific Tableland and Kisbey fields.

Unfortunately Veritas Software broke up shortly afterwards, the Vista software became the property of one of the emerging companies, Seismic Image Software, and I had to say goodbye to my first, and very own, personal computer workstation.

The fine art of Mother Nature

Chris Kent

With sequence stratigraphic concepts and a modicum of understanding of the regional tectonic history, your 3D seismic cube can begin revealing its secrets. Before we delve into this further, it's important to remember that by far the most gracious method of understanding what comes out of the data, is to look at the world around us. All the depositional environments that have ever existed still exist today; as expounded by Charles Lyell in the early 1830s, the present is the key to the past.

In this sense, imagination is more important than knowledge (a quote from Einstein), and for those geophysicists who do not spend enough time in the real world, perpetually locked in the office, you must venture into Google Earth for a cut-price field trip. If you want to see some fluvio-glacial outwash channels, just go to Alaska. If you want to go see a nearshore bar, the west coast of Denmark is awesome:

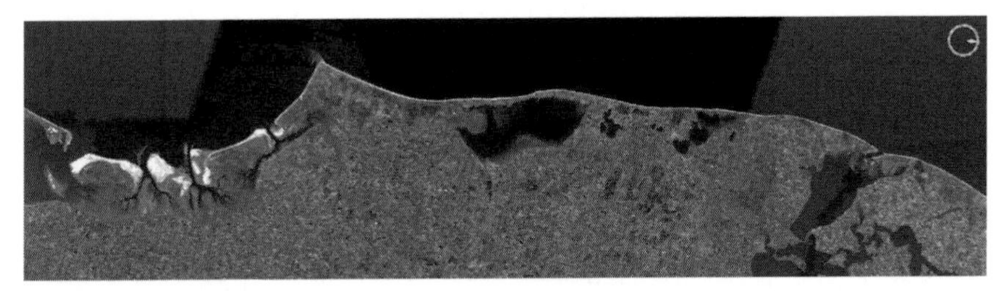

Don't forget nature, don't forget scale, and keep it real. Nature has given you a cheat sheet to the subsurface exam!

With this in mind, you're now armed with a dangerous skill set. Once you've got some sequence boundaries mapped out, you can start the detective work. Thickness variation and seismic truncations, once you have accounted for structural elements, will guide you to deductions about relative sea-level. Push the seismic data to its limits; forget about specific attributes, just run as many as you can, because you may see something interesting, and even the absence of something interesting is information in itself! Give your data an MRI scan: run a spectral decomposition tool which can unlock hidden information. Target your analysis and try not to cut across too many onsets, you need to be parallel to reflections as much as possible to visualize bedform patterns on palaeoslopes.

Don't forget nature, don't forget scale, and keep it real. Nature has given you a cheat sheet to the subsurface exam!

Remember to use unfiltered, high-fidelity data where possible, and be realistic when it comes to your stratigraphy. For example, if you have a hard chalk sequence above your target zone, you can kiss some of the high frequency content goodbye. If you can, co-rendering attributes in a volumetric visualization will give a big advantage.

Immerse yourself into the geology that shape-shifts the world around us. Whenever you are on a beach, by a river, or flying 10 000 metres above the ground, you can gain increased understanding of the subsurface. It's so easy to lose sight of what you're working on in the office and field trips may be infrequent. So, take five, and Google away.

The idea of seismic attributes

Art Barnes

Seismic attributes are tools for inferring geology from seismic reflection data. Seismic attributes aid seismic interpretation by revealing subtle features, by identifying similar patterns, and by quantifying specific properties. Attribute analysis is a vital facet of reflection seismology for petroleum exploration and finds application from anomaly identification to feature extraction to lithologic prediction.

Seismic attributes are quantifiable properties of seismic data. They are subsets of the information in the data, and in this way simplify data interpretation. Attribute computations resemble data processing methods, but there is a distinction. In data processing, the goal is to enhance the signal by removing noise. In attribute computation, the goal is to enhance features of interest by removing some portion of the signal.

Attribute analysis decomposes data into attributes. The decomposition is informal; no rules govern how to compute attributes or what they must represent. In effect, attribute computations are filters that remove some portion of the data to reveal hidden features of interest, such as bright spots, faults, and channels. It is often argued that seismic attributes are never as good as the original seismic data because they have less information. This criticism misses the mark entirely — attributes are useful precisely *because* they have less information.

Seismic attributes are applied to pre-stack data gathers or post-stack data volumes. Pre-stack attributes measure amplitude changes and derived rock properties such as compressional and shear velocities or impedances. Post-stack attributes measure amplitude, frequency, discontinuity, dip, parallelism, and waveform, among others. Pre-stack attributes treat seismic data as recordings of seismic reflections. Post-stack attributes treat seismic data as images of the earth. Pre-stack attributes are derived through involved methods of geophysical inversion. They provide valuable clues about lithology and fluid content, but they are relatively expensive, demand careful interpretation, and require sophisticated data preparation. Post-stack attributes are derived through filters, transforms, and statistics. They quantify stratigraphic and structural properties and are easy to compute and apply, but they lack the direct ties to lithology and fluids that are of paramount interest.

Seismic data has many properties, and each property can be quantified in various ways. Hundreds of seismic attributes have been invented and more appear each year. Their great number and diversity is confusing and inhibits their application. But most seismic attributes are duplicates or unstable or lack useful meaning; they can be discarded. Discarding unneeded attributes leaves a much smaller and more manageable set of attributes that are relatively unique, stable, and meaningful. Above all, attributes should be meaningful, and preferably measure a property that is clearly related to geology or geophysics.

The two most important post-stack seismic attributes are reflection strength and discontinuity. Other useful attributes include maximum amplitude, instantaneous phase, average frequency, most positive and most negative curvature, spectral decomposition, waveform, relative acoustic impedance, and relative amplitude change. The two most important pre-stack attributes are compressional and shear impedances. Their information is often recast as Lamé's parameters, lambda-rho and mu-rho.

Here's my list of the seismic attributes that are suitable for application to key objectives in seismic data analysis.

- **Reconnaissance:** reflection strength, discontinuity, relative acoustic impedance, shaded relief.
- **Amplitude anomalies:** reflection strength, relative acoustic impedance, acoustic impedance, shear impedance, lambda-rho, mu-rho.
- **Frequency shadows:** average frequency, bandwidth, quality factor.
- **Faults:** discontinuity, most positive curvature, most negative curvature, dip, relative amplitude change, shaded relief.
- **Channels:** reflection strength, discontinuity, spectral decomposition, tuning frequency, waveform, acoustic impedance.
- **Stratigraphy:** instantaneous phase, reflection strength, parallelism, average frequency, waveform.

Seismic attributes are invaluable for mapping faults and channels, and for identifying bright spots and frequency anomalies. Further, they provide a basis for geobody detection, and aid data reconnaissance and presentation. Attribute interpretation remains largely a matter of qualitative investigations with individual attributes, but quantitative multi-attribute analysis promises greater rewards. However, current methods of multi-attribute analysis remain inadequate and must be improved greatly if we are to further automate aspects of data analysis. Therein lies the challenge for the future of seismic attributes.

The last fifteen years

Dave Mackidd

There are only two things in the universe that remain constant over time:

1. Change.
2. Resistance to change.

Perhaps the biggest contribution to interpretative geophysics in the last decade and a half has been Bill Goodway's introduction of the lambda-mu-rho or LMR formulation for amplitude-versus-offset (AVO) inversion. Dave Cooper likes to say that LMR really stands for 'Let's Measure Rocks'. But as with all great ideas, Bill encountered his share of detractors. I used to tell him, 'That's proof you are on the right track!' The naysayers argued that although it opens up the crossplots and makes them easier to interpret, LMR is no more precise than any other AVO method. They failed to recognize the significance of the method.

Every geologist is trained from the outset to differentiate rocks based on their hardness and brittleness, and that's precisely what lambda and mu are. Few geologists have an intuitive feeling for Poisson's ratio or Young's modulus. Bill had formulated AVO in a framework which transcended geophysics, connecting to the disciplines of geology and engineering. More importantly, managers could now understand the AVO crossplots! LMR also removed the difficulty of reconciling static and unbounded laboratory measurements with the dynamic bounded measurements of seismic waves. As Bill has demonstrated many times, most standard equations in anisotropy and geomechanics add more insight when they are expressed in LMR terms.

The Megabin method

Bill also recognized that pre-stack interpretation needed better 3D sampling than was provided by standard orthogonal acquisition methods. He pioneered the use of interpolation as an integrated part of acquisition to achieve better offset and azimuth sampling with his patented Megabin technique. This concept was first proposed by Jon Claerbout in a course that Bill and I took together in 1994, as well as in one of his textbooks. However, to my knowledge, Bill was the first to actually go out and do it. With the arrival of better interpolation techniques, the approach is now a common part of many wide azimuth methods.

The geophysicist of today has to be an expert in the areas of anisotropy and geomechanics, and understand new tools, such as microseismic data.

The resource play upheaval

It's not only new tools that have made the last 15 years so exciting. The emergence of the resource play concept has brought about vast changes in interpretive geophysics. Advances in technology are making accessible the huge volumes of hydrocarbons that reside in the tighter reservoirs of the world (the lower end of the resource pyramid). This idea was pioneered by Canadian Hunter in the Deep Basin in the 1970s and by Encana everywhere in North America in the past decade. Almost overnight, seismic interpretation went from being the leading edge of exploration to what some viewed as a lesser development support role. In its wake, this transformation left many a bitter and disillusioned geophysicist, not the least of whom was Bill Goodway. That is until Bill snapped out of it and adapted to the new reality. Now, once again, he is leading the way for the rest of us. You see, it turns out that geology is not one of the universal constants, something the 'gas factory' models tend to overlook. There is still a major role for us to play, but it has evolved. The geophysicist of today has to be an expert in the areas of anisotropy and geomechanics, and understand new tools, such as microseismic data.

If there is a lesson for a young geophysicist, it is this: your most important career skill is the ability to adapt to change, instead of resisting it. And this is becoming even more important as we wake up to the third universal constant:

3. The rate of change is increasing!

I have had an extremely satisfying and exciting career. I have walked among and stood upon the shoulders of giants: Claerbout, Robinson, Treitel, Taner, Hampson, Russell, Goodway, to name just a few! However, when I go through that door for the last time, my final thought will be, 'I wonder what geophysics will be like 40 years from now.'

I guess I'll have to hang around and find out…

The magic of Fourier

Mostafa Naghizadeh

If I were to respond to the question 'Which single mathematical idea had the greatest contribution to the modern world of technology?' I would reply, without any hesitation, 'The Fourier transform'. No one can deny the great role that the Fourier transform has played in telecommunications, the electronics industry, and signal-processing methods over the past century. Therefore, it is no surprise that the Fourier transform also has an indispensable part to play in geophysical data analysis. While the presence of Fourier can be felt in every discipline of geophysics, here are some highlights from exploration seismology:

Data acquisition. Almost all of the electronic and mechanical devices used in field data acquisition are designed and tested using the Fourier analysis. Also the collected signals have to be checked by Fourier analysis to ensure acquisition of a proper bandwidth of data. Bandwidth has a direct relationship with the strength and resolution of a signal; it's a good measure of a signal's capacity to obtain information from desired depths of the subsurface. The Fourier transform also plays a prominent part in data compression and storage.

Noise elimination. The collected seismic data are often contaminated with random or coherent noises. In the traditional time-space coordinates that the seismic data are recorded in, one can not easily distinguish between noise and signal. However, by transforming the data to frequency and wave-number domains using the Fourier transform a significant portion of noise is separated from the signal and can be eliminated easily.

Interpolation. Seismic records contain specific features of simplicity such as a stable wavelet and spatial continuity of events. This simplicity has even more interesting consequences in the Fourier domain. Seismic data composed of linear events will only have a few non-zero wave-number coefficients at each frequency. This sparsity, as it is called, is used in most of the seismic data interpolation methods to produce a regular spatial grid from irregular spatial sampling scenarios. Also, a simple relationship between the low frequencies and high frequencies of the seismic records helps to overcome the spatial aliasing issues by reducing spatial sampling intervals.

Time-frequency analysis. Seismic traces can exhibit non-stationary proper-

No one can deny the great role that the Fourier transform has played in telecommunications, the electronics industry, and signal processing methods over the past century.

ties. In other words, the frequency content of the data varies at different arrival times. Likewise, spatially curved seismic events have non-stationary wave-number content. In order to analyse these signals, time-frequency analysis is required. Several transforms such as Wavelet, Gabor, Stockwell, and Curvelet transforms have been introduced to perform these analyses. In essence, all of these transforms are windowed inverse Fourier transforms applied on the Fourier representation of the original data. These Fourier-based transforms can be used for interpolation and noise elimination of very complex seismic records.

Migration and imaging. Migration of seismic records is a necessary processing step to obtain the true depth and dip information of seismic reflections. The Fourier transform is a very common mathematical tool for solving wave-equation problems. Many of the wave-equation migration techniques deploy the multidimensional fast Fourier transforms inside a least-squares fitting algorithm for imaging the subsurface. The Fourier-based wave-equation migration algorithms avoid the common pitfalls of ray-tracing migration algorithms known as ray-crossing.

The magic of Lamé

Bill Goodway

If geophysics requires mathematics for its treatment,
it is the earth that is responsible not the geophysicist.
Sir Harold Jeffreys

This quote was offered as a disclaimer on a course I took at the University of Calgary in 1988: Dr Ed Krebes' Geophysics 551 *Seismic Techniques*. This excellent course was pivotal in my enlightenment regarding Lamé's parameters. I repeat the quote here as it disclaims my seemingly unnecessary obfuscation in the use of equations that follow.

The basic earth parameters in reflection seismology are P-wave velocity V_P, and S-wave velocity V_S. However, these extrinsic dynamic quantities are composed of the more intrinsic rock parameters of density and two moduli terms, lambda (λ) and mu (μ), introduced by the 18th-century French engineer, mathematician, and elastician Gabriel Lamé. Lamé also formulated the modern version of Hooke's law relating stress to strain as shown here in its most general tensor form:

$$\sigma_{ij} = c_{ijkl}\,\varepsilon_{kl} = (\lambda\delta_{ij}\,\delta_{kl} + \mu\delta_{ik}\,\delta_{jl} + \mu\delta_{il}\,\delta_{jk})\,\varepsilon_{kl}$$

Here, σ_{ij} is the i-th component of stress on the j-th face of an infinitesimally small elastic cube, c_{ijkl} is the fourth rank stiffness tensor describing the elasticity of material, ε_{kl} is the strain, and δ_{ij} is the Kronecker delta. The adage 'stress is proportional to strain' was first stated by Hooke in a Latin anagram *ceiiinosssttuv*, whose solution he published in 1678 as *Ut tensio, sic vis* meaning 'As the extension [strain], so the force [stress].' Despite being interestingly reversed and non-physical, Hooke's pronouncement is illustrated here with complete mathematical rigor, and this equation creates the basis for the science of materials, including rocks. Interestingly, and most notably, only Lamé's moduli λ and μ, appear in this equation; not bulk modulus, Young's modulus, Poisson's ratio, V_P, V_S, or any other seismically derived attribute.

The methods to extract measurements of rocks and fluids from seismic amplitudes are based on the physics used to derive propagation velocity. This derivation starts with Hooke's law and Newton's second law of motion, and yields a set of partial differential equations that describe the progression of a seismic

The methods to extract measurements of rocks and fluids from seismic amplitudes are based on the physics used to derive propagation velocity.

wave through a medium. It also forms the basis of AVO-based descriptions of the propagating medium.

The P-wave propagation of a volume dilatation term θ derived from Hooke's law is:

$$\rho \frac{\partial^2 \theta}{\partial t^2} = (\lambda + 2\mu) \nabla^2 \theta$$

and the S-wave propagation of the shear displacement term $(\nabla \times u_{sh})$:

$$\rho \frac{\partial^2 (\nabla \times u_{sh})}{\partial t^2} = \mu \nabla^2 (\nabla \times u_{sh})$$

The vector calculus in these equations says that the particle or volume displacement for a travelling P-wave in the earth is parallel to the propagation direction (as $\nabla \times \theta = 0$), whereas the particle displacement imposed by a passing S-wave is orthogonal to the travelling wavefront. Consequently, the intuitively simple Lamé moduli of rigidity μ and incompressibility λ afford the most fundamental and orthogonal parameterization in our use of elastic seismic waves, thereby enabling the best basis from which to extract information about rocks within the earth.

These Lamé moduli form the foundation for linking many fields of seismological investigation at different scales. Unfortunately the historical development of these fields has led to the use of a wide and confusing array of parameters, which are usually complicated functions of the Lamé moduli. None of these are inherent consequences of the wave equation, as the Lamé moduli are. This includes standard AVO attributes such as intercept and gradient or P-wave and S-wave reflectivity that are ambiguous and complex permutations of Lamé moduli λ and μ, or Lamé impedances $\lambda\rho$ and $\mu\rho$. Many other parameters such as Poisson's ratio and Young's modulus have arisen due to inappropriate attempts to merge the static un-bound domain of geomechanics to the dynamic bound medium of wave propagation in the earth. These attempts have resulted in the use of contradictory assumptions, which are completely removed when restating equations using the magic of Lamé moduli.

The scale of a wavelet

Brian Romans

It doesn't take long to get accustomed to the ability to sit comfortably at a computer workstation and cruise around a subterranean world with just a few mouse clicks. The technology we use to observe, describe, and interpret subsurface geology is truly amazing. The advent of 3D seismic-reflection data coupled with immersive visualization software allows us to characterize and, importantly, to conceptualize the heterogeneity and dynamics of reservoirs.

With this power, it's sometimes easy to forget the scale of geology we are dealing with in subsurface data. The scale of features combined with the type and resolution of data you are looking at can often lead to interpretations that do not capture the true complexity and heterogeneity.

I find it useful to constantly ask myself questions about scale when interpreting the subsurface: How thick is the package of interest? How wide is it? Take a few minutes and do the back-of-the-envelope calculations to see how many Empire State buildings (about 1 million cubic metres) or Calgary Saddledomes (250 000 cubic metres) fit inside your volume of interest.

When you figure out display settings that best work for you and the data you are characterizing, calculate the vertical exaggeration. Write this on a sticky

Do the back-of-the-envelope calculations to see how many Empire State buildings or Calgary Saddledomes fit inside your volume of interest.

note and attach it to your monitor. It's quite common to view these data at high vertical exaggerations — especially in fields where subtle stratigraphic traps are the issue.

Finally, and most importantly, go on a field trip at least once every couple of years. Observing and pondering geology in the field has numerous intellectual benefits — far too many to list here. Seek out the extraordinary, like the turbidites onlapping the basin margin in the Eocene of the French Alps opposite. The realization of scale is among the most critical. Spending an hour or more trekking your way up a rocky slope to see some outcrops and then finding out all that expended energy and sweat got you through about half a wavelet is an unforgettable experience.

Acknowledgments

Image by the author. See *ageo.co/LbQLUg* for the full story.

The subtle effect of attenuation

Fereidoon Vasheghani

When you wake up in the morning, you look at yourself in the mirror — at least I hope you do! You are able to see yourself because light rays are reflected when they hit the opaque but shiny back surface of the mirror. Later, when you are on a bus or train, you see yourself reflected in the window, but not as sharply as in the mirror. That is because only a small percentage of light rays reflect back; most of them transmit through the glass. That is why others outside the bus can see you.

The same principles apply to the seismic waves in the earth. The following famous equation for reflection coefficient R shows how much of the wave amplitude is reflected from a normally incident seismic ray when it hits a boundary:

$$R = \frac{\rho_2 V_2 - \rho_1 V_1}{\rho_2 V_2 + \rho_1 V_1}$$

where ρ and V are density and P-wave velocity, respectively. The subscripts represent the layers across the boundary, with 1 being the upper layer. The product of the velocity and density is called the acoustic impedance. The equation shows that when there is an acoustic impedance contrast between the two layers, there will be reflection. But this equation is actually an approximation. Research shows that there is another contrast that contributes to the reflection of the waves and rays in exploration seismology, and that is attenuation contrast.

Attenuation is the loss of energy in the form of heat. Particles oscillate in their place without flowing from one location to another, to allow the seismic waves to go through. These particles turn some of the seismic wave energy into heat due to friction forces. You can think of this energy loss as a sort of currency exchange fee. Every time you exchange one currency for another, the bank charges a transaction fee, which is deducted from your money. If you do the exchange several times, you lose more and more of your money. As an experiment, start with some of your local currency and exchange it to a few foreign currencies in turn and then convert it back to your local money. You will see that a significant portion of the money is gone! This loss, in seismic terms, is called attenuation and is measured through a parameter called quality factor Q. In fact Q is inversely proportional to the attenuation.

Research shows that there is another contrast that contributes to the reflection of the waves and rays in exploration seismology, and that is attenuation contrast.

When we account for attenuation as well as acoustic impedance, Lines et al. (2008) showed that we can write a more accurate version of the reflection coefficient equation:

$$R = \frac{\rho_2 V_2 \left(1 + \frac{i}{2Q_2}\right) - \rho_1 V_1 \left(1 + \frac{i}{2Q_1}\right)}{\rho_2 V_2 \left(1 + \frac{i}{2Q_2}\right) + \rho_1 V_1 \left(1 + \frac{i}{2Q_1}\right)}$$

It results in a complex reflection coefficient. This equation reduces to the simpler form when there is no contrast in the quality factors, i.e. when $Q_1 = Q_2$.

Imagine a two layer model with the following properties: $\rho_1 = \rho_2$, $V_1 = 2000$ m/s, $V_2 = 3500$ m/s, $Q_1 = 40$, and $Q_2 = 6.28$. The reflection coefficient R calculated using the simple equation opposite is 0.273, which means the amplitude of the reflected wave is 0.273 times the amplitude of the incident wave. The result including the attenuation effect is the complex number $(0.274 + 0.031i)$, which means the reflected amplitude is 0.276 times that of the incident wave (0.276 is the *modulus* or magnitude of the complex reflection coefficient), and the phase rotation between the incident and reflected waves is 6.4° (the *argument* or phase angle of the complex number).

Since the difference between the predicted amplitudes and phases of the reflected waves calculated from two equations is small (in real cases, seismic noise makes the relative effect even smaller), it is within reason to say that the familiar equation is a good approximation of the reflection coefficient for most scenarios. But Q is in there somewhere!

References

Lines, L, F Vasheghani, and S Treitel (2008). Reflections on Q. CSEG *Recorder* **33**, 36–38.

The unified AVO equation

Rob Simm

AVO theory provides us with the mechanics for optimizing fluid and rock content from seismic data. Arguably, the most important contributions in the literature to the understanding of AVO in the last 15 years have emanated from BP. I would claim (and I must stress that I have never worked for BP, even as a consultant) that what has been presented is essentially a unified view of AVO, in as much as it seems to explain many historical observations about AVO and puts into perspective approaches presented by other authors. It is interesting in the light of this claim, however, that there are many interpreters who are ignorant of what it is all about. Maybe this is due to the academic tone of the papers.

Central to AVO is the relationship between impedance AI and reflectivity R, such that $R \approx 0.5\Delta\ln(\mathrm{AI})$. As an aid to inverting non-normal-incidence angle stacks Connolly (1999), using the Aki–Richards (1981) simplifications of the Zoeppritz equations, derived impedance at an incidence angle (elastic impedance or EI). Thus, two-term elastic impedance is effectively the integration of Shuey reflectivity (i.e. $R = R_0 + G\sin^2\theta$).

From the perspective of the AVO (intercept *vs* gradient) crossplot, Whitcombe et al. (2002) realised that Shuey reflectivity is described by a coordinate rotation or projection and that the incidence angle θ is related to the crossplot angle of rotation χ by $\sin^2\theta = \tan\chi$. In other words, the x-axis (intercept) is the axis onto which the crossplot points are projected, and as the axes are rotated in a counter-clockwise direction, projections onto the rotated x-axis describe reflectivity at increasing incidence angles. Of course these projections will look similar to equivalent seismic angle stacks only if the seismic conforms to the linear approximation.

One problem, however, is that Shuey reflectivity can only be applied over a certain range of angles. In terms of the incidence angle θ the limitation is simply $\theta = 0° – 90°$ ($\chi = 0° – 45°$). If $\sin^2\theta$ is replaced with $\tan\chi$ in Shuey's equation then at high χ angles the projected reflectivity can give values greater than unity. So, a modification to Shuey's equation is required to enable projections at any crossplot angle, giving a reasonable range of values whilst maintaining the correct relative differences between AVO responses. The result is

*This is **the** AVO equation and like most really useful equations it is elegant in its simplicity.*

$$R = R_0 \cos \chi + G \sin \chi$$

which is effectively Shuey's equation written in terms of χ and scaled by $\cos \chi$. This is *the* AVO equation and like most really useful equations it is elegant in its simplicity. Given that it effectively extends the angular range of Shuey's equation, the corresponding impedance is termed extended elastic impedance (EEI). All of this is not very sexy but it does turn out to be very useful. A few aspects are worthy of note:

- Uncalibrated AVO analysis is essentially an angle scanning operation in which the interpreter tries to identify both fluid and rock effects within the context of a regional geological model.
- The EEI formulation is useful for detailed analyses of the intrinsic AVO signature simply through crossplot analysis of log data. Introducing gradient impedance GI as the EEI at $\chi = 90°$, the AI *vs* GI — or rather ln(AI) *vs* ln(GI) — crossplot enables the plotting of all lithological units and maintains the angular relations of the intercept *vs* gradient crossplot, thus ensuring an appreciation of the linear discrimination of fluid and rock type. This analysis should be done in conjunction with bandlimited impedance synthetics to appreciate the nature of seismic resolution. It should also be remembered that theoretical fluid angles are generally upper angle limits; owing to the effects of noise, the effective fluid angle is usually lower.
- Angle-independent elastic parameters (Poisson's ratio, acoustic impedance, λ, μ, and so on) can be shown to correlate with a particular angular rotation or value of EEI.

References

Aki, K and P Richards (1980). *Quantitative Seismology, Theory and Methods.* San Francisco: Freeman.

Connolly, P (1999). Elastic impedance. *The Leading Edge* **18**, 438–52.

Whitcombe, D, P Connolly, R Reagan, and T Redshaw (2002). Extended elastic impedance for fluid and lithology prediction. *Geophysics* **67**, 63–67.

Use names to manage data

Don Herron

The sheer volume of seismic data to be evaluated and folded into fully integrated interpretations has increased the burden already carried by interpreters. We must work ever more efficiently and assess the quality of a rapidly growing number of interpretation products, many of which are the output of automated processes. To keep from being overwhelmed by a torrent of seismic datasets and derivative products (horizons, faults, grids, attributes, etc.), you must effectively manage your data. Begin by designing and using a nomenclature system in every interpretation project.

A good data-management system serves as a road map, with which a user should be able to navigate through an interpretation project with a minimum of difficulty and confusion. A system should be comprehensive, in the sense that it contains and succinctly describes all of the fundamental elements that a user needs to find a desired product. It also should be intuitive, not based on jargon or esoteric language, and flexible enough that it doesn't require re-engineering to incorporate a new data or product type. An effective data management system is one that you should carefully design at the outset of a project to facilitate rather than encumber interpretation. A good way of thinking about such a system is to imagine what you would need to know about a project if you were taking it over from another interpreter — how would you find what she had done, and determine how she had done it? (I assume, of course, that as a normal business process interpreters document their work in a report of some kind that resides in some sort of document management system.)

Let's consider a horizon naming system. Each horizon name must have a minimum number of core elements, each delineated by an underscore. These elements must appear in the following order in each unique horizon name:

- Project identifier — perhaps including the vintage of seismic data
- Numerical designation for the approximate geological age of the horizon. This number increases with the age of the interpreted horizon, and enables the horizon list to be sorted and displayed in order of horizon age.
- Colour assigned to the horizon
- Biostratigraphic age of the horizon

- Name of the trace data file on which the horizon was interpreted. This is especially important if there are, say, pre-stack and post-stack migrations
- Initials of the interpreter who picked the horizon

Here's an example, with an explanation of what all the pieces mean:

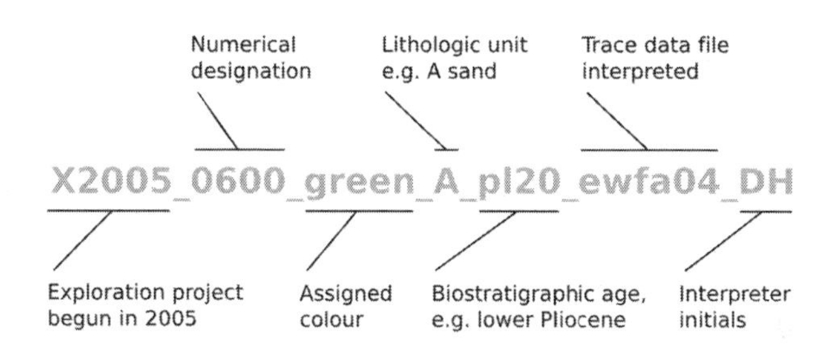

X2005	Exploration project that began in the year 2005
0600	Numerical designation for the horizon
green	Colour assigned to the interpreted horizon
A	Indicates a named lithologic unit (in this case, the so-called 'A' sand)
pl20	Biostratigraphic age of the horizon (in this case lower Pliocene)
ewfa04	Name of trace data file on which the horizon was interpreted (in this case the final gained version of a pre-stack time migration volume with a sample interval of 4 ms)
DH	Interpreter's initials

Continuing with the road map analogy, this description of the core elements of the system should serve as the map's legend, clearly explaining the meaning of each element. This legend should be kept as an integral part of the project's documentation, perhaps as a *readme* file.

Many companies maintain nomenclature systems for their seismic projects, but even with a company-wide system in place you should always be sure to document instances in which you customize a system for the specific purposes of a given project. One or two frustrating instances of losing valuable data is enough to convince most interpreters that data management matters. Indeed, business success depends on it.

References

Herron, D A (2001). Problems with too much data. *The Leading Edge* **20**, 1124–26.

Herron, D A (2011). *First Steps in Seismic Interpretation*: SEG Geophysical Monograph Series No. **16**.

Use the rock physics bridge

Per Avseth

Rock physics represents the link between geology and geophysics. Increasingly over the last decade, rock physics stands out as a key technology in petroleum geophysics, as it has become an integral part of quantitative interpretation of seismic and electromagnetic data. Ultimately, the application of rock physics tools can reduce exploration risk and improve reservoir forecasting in the petroleum industry.

Mind the gap

Traditionally, rock physics has focused on the understanding of how elastic properties and seismic signatures change as a function of hydrocarbon saturation, porosity, and pressure. With great breakthroughs in laboratory experiments and theoretical modelling, rock physics has extended its turf and today plays an important role in the basin scale characterization of the subsurface, being an integral part of well log, seismic, and electromagnetic data analysis.

The role of rock physics as a bridge between geology and geophysics poses new challenges and opportunities. The introduction of rock physics 'templates' (Ødegaard and Avseth 2004) as a tool for interpretation and communication has proven beneficial to the oil industry. The nifty thing with the templates is that rock physics models can be constrained by local knowledge from experienced geologists. Furthermore, the templates force the geological interpretation of well log and seismic data to be made within rigorous physical bounds. We can also use the rock physics templates to extrapolate away from a few observed wells in the earth and say something about expected rock properties and seismic signatures for various lithology and pore-fluid scenarios.

The sound of geology

Recent research studies have highlighted the importance and benefit of linking rock physics to geologic processes, including depositional and diagenetic trends (e.g. Dræge et al. 2006, Avseth et al. 2010). These studies have proven that lithology substitution can be as important as fluid substitution during seismic reservoir prediction. It is important during exploration and appraisal to extrapolate away from existing wells, taking into account how the depositional environment changes as well as burial depth trends. In this way rock physics can better constrain the geophysical inversion and classification problem in underexplored marginal fields, surrounding satellite areas, or in new frontiers.

It turns out that the rock texture we observe in cores and thin sections at the microscale strongly affects the seismic reflection amplitudes that we observe at the scale of tens of metres. We can apply rock physics templates to interpret sonic well-log measurements or seismic inversion data. In a way, we are using our effective rock physics models to downscale our geophysical observations to geological information about subsurface rock and fluid properties.

The memory of rocks

It is important to honour not only the present-day geology when we use rock physics templates for geological interpretation of well and seismic data. We should also know the burial history of the rocks. The rocks have 'memory' of the stress and temperature history since deposition, from mechanical and chemical compaction to episodes of perhaps uplift and burial. Therefore we occasionally observe well-cemented and high velocity rocks not corresponding with present-day temperatures and depths.

In other words, it is important to take into account the memory of rocks, as temperature and stress history make a significant imprint on reservoir and seal rocks. This is particularly important in areas with complex tectonics and uplift. With a better integration of basin analysis and geophysical interpretation, via rock physics models, we can more reliably interpret lithology and fluid parameters during hydrocarbon exploration and production.

Let's rock it!

In a world in which energy insecurity is at the forefront of global challenges, building bridges across disciplines is a requirement for new discoveries and improved oil recovery. The field of rock physics has evolved to become one of these bridges, bringing geology and geophysics closer together. The time when large gaps separated our earth science disciplines is definitely over. So let's rock the future together!

References

Avseth, P, T Mukerji, G Mavko, and J Dvorkin (2010). Rock-physics diagnostics of depositional texture, diagenetic alterations, and reservoir heterogeneity in high-porosity siliciclastic sediments and rocks — A review of selected models and suggested work flows. *Geophysics* **75**, 7531–47.

Dræge, A, T A Johansen, I Brevik, and C Thorsen Dræge (2006). A strategy for modeling the diagenetic evolution of seismic properties in sandstones. *Petroleum Geoscience* **12** (4), 309–323.

Ødegaard, E, and P Avseth (2004). Well log and seismic data analysis using rock physics templates. *First Break* **22**, 37–43.

We need integrative innovation

Maitri Erwin

A big challenge for geophysicists and seismic interpreters is to do good work today and to make results available to tomorrow's workers. With rapidly-advancing subsurface data acquisition, analysis, storage, and access tools at our disposal, how do we maximize their use for accurate and efficient project execution and retain information during and after project completion? This requires a two-fold approach: geoscience-focused technological innovation, and a philosophy of interdisciplinary collaboration and communication.

Technology fit for purpose

Seismic analysis is a synthesis of many different data types — the seismic data itself, wells, velocity models, geology, cores, previous work — which in turn generates more data and parameters that are important to reservoir engineers, economists and drillers. Furthermore, geoscientists are inherently interested in exchanging ideas and information, but not necessarily the information-technology business models whereby that happens.

What will be helpful then are not new software applications in which to create, analyse, and record, but versatile innovations that help us increase the efficiency of existing data and tools required for interpretation, archival, and transmission. These solutions will acknowledge and fit the subsurface workflow of geophysical interpretation and model building, followed by dynamic well planning and drilltime model refinement, and not the other way around. The user should not be constrained by the tools.

Work together and bridge the gap

The geoscience community has the same problem as the intelligence community. Each person on the project has at least one crucial bit of information that no one else possesses. Analyses also create an immense wealth of knowledge that is not effectively transmitted through the organization and to the next generation of workers. Technical proficiency in geophysics and interpretation is traditionally rewarded, and collaborating with geologists, petrophysicists, reservoir engineers, economists, and drillers takes time. Such collaboration is invaluable as we move forward, however, especially in the areas of inversion and 4D interpretation which involve reservoir architecture,

What will be helpful then are...
versatile innovations that help us increase
the efficiency of existing data and tools.

volumetric properties, and fluid movement. Adopted at a management level, a culture of sharing within projects will encourage similar interdisciplinary associations across the organization.

As scientists in a cutting-edge industry, we know that technology is not just software but also the effective utilization and management of data and people. Flexible and light solutions that address our workflow requirements and boost the capabilities of existing tools as well as cross-disciplinary work and communication are two ways to get us there.

Well tie basics

Rachel Newrick

Borehole measurements such as gamma ray, resistivity, sonic and density logs, rock cuttings, core samples, casing points, and biostratigraphic results are all recorded in depth. Conversely, seismic measurements and our interpretations are inherently in seismic travel time.

To convert borehole measurements from depth to time, or to convert the seismic interpretation from time to depth, a time–depth relationship needs to be established. We can use one of many techniques including velocities derived from seismic, a checkshot or VSP, a sonic log, or a combination of any of these. We confirm the time–depth relationship at the borehole location by generating a synthetic seismogram.

To make a synthetic seismogram we need to:

1. Generate a reflectivity series.
2. Apply a time–depth relationship.
3. Convolve with a wavelet and compare to the seismic data.

If the synthetic seismogram is a good match to the seismic we can say that the time–depth relationship is robust, and that the borehole data are located accurately on the seismic section and can be confidently extrapolated outwards.

Generating a reflectivity series

Reflectivity is generated by changes of impedance $I = \rho V_P$ within the earth. Since impedance is the product of velocity (V_P) and density (ρ) we can generate a reflectivity series directly from the slowness (DT) and bulk density (RHOB) curves. A full suite of quality velocity and density logs is not always available, so pseudo-curves can be estimated using established relationships like Faust or Gardner, as discussed in *Well tie perfection*.

Estimating the time–depth relationship

We use all the information available to us to generate the time–depth relationship — remember, it is all about the time–depth relationship. Commonly, we start by integrating the sonic log to estimate time–depth pairs, that is, we sum up all the measurements to get a total travel time to each depth in the bore-

A full suite of quality velocity and density logs is not always available so pseudo-curves can be estimated using established relationships like Faust or Gardner.

hole. Because sonic velocities are not the same as seismic velocities, due to the phenomenon called dispersion, and because there are often gaps and spurious measurements in the sonic log, the integrated sonic velocities often leave an incomplete record that provides a poor tie. We can calibrate the sonic velocities with a checkshot survey.

The checkshot survey is a measurement of seismic travel time at a range of depths in the borehole, at least at key stratigraphic boundaries and total depth. With checkshot data, we are saying, in effect, that we know how long it takes for seismic energy to travel to this depth. So the time–depth relationship must include these points.

In a marine setting, another time–depth point is the time and depth of the seabed reflection. The seabed time can be read from seismic and the seabed depth is recorded in the well file.

Pulling it together

We convolve the reflectivity series with a wavelet to give the appearance of the seismic. Using the estimated time–depth relationship, the synthetic seismogram can be compared directly to the seismic. If there is a good set of logs, a wavelet that approximates that of the seismic section, and a good time–depth relationship we should have a good tie between the seismic and the borehole. The synthetic will be a 'good match' to the seismic, with similar frequency content, high amplitudes in the same place, transparent zones in the seismic matched by little reflectivity in the synthetic seismogram, and not much dispute from anyone who looks at the tie.

Often we are not so fortunate. I outline some ways to deal with poor ties in *Well tie perfection*.

Well tie perfection

Rachel Newrick

The beauty of modern well-tie software is that it is easy to pull in a few curves, to add some pin points and to move the synthetic to match the seismic, stretch a bit of the top, perhaps squeeze a bit of the base. So let's think about what we are actually doing when we apply these processes.

Bulk shifting

Often the entire synthetic appears to be too shallow or too deep. There is likely a discrepancy in the datum or the replacement velocity used to anchor the uppermost point of the synthetic. In this case it is valid to apply a bulk shift but always check the datum and review the interval velocities in the near surface to make sure that they are reasonable.

Stretch and squeeze

In my experience this is the most contentious part of the well-tie process because it is easy to abuse. This is not a procedure by which you simply select peaks and troughs and match them to the seismic. You can make anything tie that way.

The idea is that a distinctive reflector on the seismic section is also observed on the synthetic (and is thus observable on the borehole logs) and there is a certainty that it represents the same event. If the event is identified in the borehole as a specific unconformity that has been correlated from other wells on the seismic, so much the better.

There is well-documented dispersion between sonic and seismic velocities. Dispersion is the phenomenon of frequency dependence of acoustic velocity in the rock. We usually need to make a correction for this by reducing the sonic velocities by a small amount. This is most easily undertaken by stretching the synthetic seismogram so that the depth extent occurs over more time. All geophysicists have their own thoughts on the procedure, but I like to first slightly stretch the entire synthetic so that, in effect, a single drift correction is applied and the major reflectors are correlated.

That said, there might be a significant change of lithology (e.g. clastic to carbonate) so a single drift correction may not apply. In this case, you might need to insert some intermediate knee points.

Great match

At this point, it is good to remember that the end game is to correctly place the depth data onto the seismic section so that we can extrapolate away from the borehole or drill a prospect, for example. It is important not just to know where we are in the borehole but to be honest about how certain we are about where we are in the borehole.

Think about what you are doing

With each modification to the synthetic we should think about why we are applying a certain process, what the effect is, and whether it makes sense. Anderson and Newrick (2008) highlighted what can and does go wrong with synthetic seismograms and I add to that here:

- Quality check the logs with the help of a petrophysicist. If adequate velocity and density curves are not available, then substitute computed curves when necessary, but be clear about what you did. There are many ways to model missing data (e.g. Gardner, Faust, Smith) so ensure that the one you choose is appropriate for the geological setting. Present a series of synthetics to illustrate the substitution (i.e. the raw curves with gaps; more complete pseudo-curves) indicating where the real data are and where the computed data are.
- Acquire VSPs for a detailed time–depth relationship. This will provide both a corridor stack to match to the seismic and synthetic, and a time–depth relationship. To extract checkshots, select some key impedance boundaries and use those time–depth pairs.
- Check that the interval velocities are reasonable for the expected strata in the borehole, and if they are not, find out why. It could equally be a data problem or interesting geological information.
- Always make a few synthetics with a variety of edited curves, different wavelets, and even different seismic volumes, such as full, near, mid, and far stacks.
- Remember that a poor synthetic tie is not always caused by poor synthetic inputs — the seismic data may not be optimally processed or correctly positioned due to many reasons, including strong velocity variations, anisotropy, or complex structure.
- When the tie just doesn't seem to work consider the amplitudes and zones of interest, i.e. the dim zone generally ties to the low amplitude zone on the synthetic, but in this case be honest about how good the tie is.

References

Anderson, P, and R Newrick (2008). Strange but true stories of synthetic seismograms. CSEG *Recorder* **33** (10), 51. Available online at *ageo.co/HZdznN*.

What I learned as a geophysicist wannabe

Victoria French

When I was an undergraduate my favourite professor, Dr Oliver T Hayward, gave me what I thought was the most challenging field project of the entire class. I was to traipse across Texas collecting data on High Plains Gravels. This was not the first challenge he had given me for his class assignments: another gem had been to determine the herding behaviour of dinosaurs. To put this in context, this was before you could search online and find oodles of information on almost any topic.

Approaching the assignment of High Plains Gravels with a less-than-enthusiastic attitude, I set out to look at every knoll and plateau across the western plains of Texas, spending a fortune on gas and cursing Dr Hayward every mile of the way. Slowly, inevitably, my attitude changed. I learned that there had been a complete inversion of topography — low areas had accumulated quartzite gravel and then resisted erosion.

I became a detective, solving a geological mystery. There is no source for the quartzite gravels for hundreds of miles, so where did they come from? How did the gravels travel so far? Why had they resisted erosion so well? It was a fascinating project from so many perspectives.

You may wonder what this has to do with being a geophysicist wannabe. The answer is that I learned the critical importance of looking for all the answers to a problem, wherever they might be, and the importance of not limiting myself by selective thinking. There are so many geological puzzles that can be better defined and answered through geophysical interpretation. Or more precisely, through the geological interpretation of geophysical data.

One of my favourite tools has been to use millisecond timeslices through merged 3D datasets to visualize basement fault reactivation and determine how it affects stratigraphy through geological time. For example, you might see how reactivated faults create carbonate margins that develop over time due to subtle highs. Perpendicular to the platform highs you may see where orthogonal fault systems have created avenues for bypass sediments shed off into the localized basins, potentially leading to a number of potential plays such as deepwater fan deposits. A case of a simple method providing profound insights.

There are so many geological puzzles that can be better defined and answered through... the geological interpretation of geophysical data.

Geophysics is such an integral part of the geological sciences, and you do not have to be a certified geophysicist to gain geological understanding by digging into geophysical data. The opportunities for solving geological problems and gaining geological insight are endless. Ask questions. Learn about data acquisition. Go to processing meetings. Read up on the new techniques. Pay attention to pitfalls. Every geologist should look at seismic data.

Where did the data come from?

Rachel Newrick

We calibrate geophysical models with physical measurements of the earth's properties. It is therefore important that we understand how the data are acquired and processed before landing on our desks for interpretation. Conversely, understanding how the data will be interpreted helps the acquisition and processing teams produce the best product.

To improve the interpretability of geophysical data, whether it's surface seismic, VSPs, well logs, gravity, magnetic, or some new technology, we should know the theory, help plan acquisition and processing, get out into the field as often as practical, quality check data processing, and help spread knowledge between teams.

Know the theory so that you can push the limits of interpretation without over-interpreting the data or falling into pitfalls. This is especially important when evaluating or applying new technologies. Ask questions, consult experts, and improve your understanding as the technology progresses.

Help plan acquisition and processing so that it is fit for purpose. Survey designs and processing parameters depend on a number of factors including regional versus focused targets, depth of investigation, structural or stratigraphic interpretation, and required resolution.

Get into the field as often as practical to understand both the general acquisition process and the specific conditions under which the data are being acquired. Each survey is slightly different and the crew rotates regularly, so seeing the field conditions and talking to people often helps you identify when things are going wrong. On a specific seismic survey, the hole in the survey may be avoiding a swamp, the geophones may be poorly planted due to freezing conditions, and the line deviation in a marine survey may be due to icebergs or existing infrastructure.

Quality check data processing so that you understand how the final measurements were derived from the field data. More importantly, add input where needed (for example, ensuring that velocities are geologically reasonable), build a rapport with the processing team, and ensure that the data are fit for purpose.

If we ensure that we at least take away good data then we have something to build the story upon and are more likely to be successful in the future.

Spread knowledge between teams so that everyone is aligned in acquiring good geotechnical measurements. This might be needed in unexpected places. Recently, I visited a frontier exploration rig that had failed to find hydrocarbons at the target. Morale was low and the crew was mentally ready to pack it in. They didn't see much use in spending time recording final logs or a VSP. Most had no geoscience background and didn't understand why we would drill in that location in the first place. I heard things like 'what on earth are we doing out here?' So, to give the project context, I gave an impromptu presentation on frontier exploration covering the identification of a basin, understanding the petroleum system, putting the geotechnical story together, and, most importantly, the value of acquiring data. I told them that if we rush the data acquisition we may walk away with nothing, but if we ensure that we at least take away good data then we have something to build the story upon and are more likely to be successful in the future. By hearing how we used the data, the crew had more ownership of the process and became part of the exploration team.

Why you care about Hashin–Shtrikman bounds

Alan J Cohen

One of the more-frequently cited and widely-used articles by geoscientists, engineers, and physicists is the classic paper by Hashin and Shtrikman entitled *A variational approach to the theory of the elastic behaviour of multiphase materials*, published in 1963. The Hashin–Shtrikman (HS) approach yields bounds (lower and upper estimates) for the elastic moduli of a homogeneous, isotropic mixture of different materials, given the elastic moduli and volume fractions of the constituents. The HS bounds on the bulk modulus have been shown to be the best possible — that is, the tightest — given only the elastic moduli and fractions of the constituents, and without specifying anything about the geometry or texture of the mixture. Although a similar proof has not been given for the bounds on the shear modulus, the HS shear modulus bounds are also believed by many authors to similarly be the best possible, and are frequently described as such.

Hashin–Shtrikman bounds can also be constructed for other physical properties, such as conductivity, effective dielectric permittivity, magnetic permeability, thermal conductivity, and so on. This follows from the mathematical equivalence of the equations that describe these phenomena. Those interested are encouraged to read, for example, Chapter 9 of *The Rock Physics Handbook* by Mavko et al. (2003), where one can also find the Hashin–Shtrikman bound equations.

Why might you care?

The HS bounds on the bulk and shear moduli have been used in a number of geoscience applications. Some authors have used them to describe the elastic moduli of mineral mixtures, towards describing the compressional and shear wave velocities in mixed-mineral sedimentary rocks, such as limestones containing both calcite and dolomite, or sandstones containing quartz and feldspar. Yet other authors have used the bounds to roughly describe the elastic moduli of unconventional resource shales by mixing clay, organic matter, quartz, calcite, and pore fluid. The elastic moduli of mudrocks, consisting of a porous clay matrix mixed with quartz, for example, are actually well described using the equations for just the lower HS bounds, as has been reported by several authors. Indeed, the mudrock geometry — consisting of an elastically weak connected clay matrix in which elastically stronger quartz is embedded — corresponds

Although the Hashin–Shtrikman bounds are, or should be, well known to geoscientists, their history is not.

well to that for which the HS lower bound is not simply a bound but an exact solution (at least for the bulk modulus).

The history of the bounds

Although the HS bounds are, or should be, well known to geoscientists, their history is not. The interested reader is encouraged to get a copy of the historical commentary by Hashin in *Citation Classics*, 1980 (available at the time of writing from *ageo.co/L7CRR5*). I offer a brief summary to whet your appetite.

Hashin and Shtrikman were two Israeli engineers who happened to be on leave in the United States at the same time. Both were involved in similar research involving predicting effective properties of composite media. Hashin was a civil engineer working on elastic properties, whereas Shtrikman was an electrical engineer working on electrical and magnetic properties. When they finished writing their seminal paper in 1961 they submitted it for publication to a prestigious American applied mechanics journal. As Hashin recalls, their manuscript was 'ignominiously rejected [by an] outstanding authority' who called their work 'ramblings.' Undaunted, the authors submitted the paper to the *Journal of Mechanics and Physics of Solids*, where it was quickly accepted, and the rest is history.

Never give up when you feel you have a good idea that should be shared with others. The geoscience community is glad Hashin and Shtrikman persevered!

References

Hashin, Z, and S Shtrikman (1963). A variational approach to the theory of the elastic behaviour of multiphase materials, *Journal of Mechanics and Physics of Solids* **11**, 127–140.

Hashin, Z (1980). Citation Classics **6**, 11 February 1980, 344. Available online at *ageo.co/L7CRR5*.

Mavko, G, T Mukerji, and J Dvorkin (2003). *The Rock Physics Handbook: Tools for Seismic Analysis in Porous Media*, Cambridge University Press.

Wrong is good for you

Bert Bril

Good geophysicists, like all good professionals, continually learn. The classic insight is that the more you know, the more you realize how little you actually know. Thus, knowledge of your ignorance grows. In my experience, lots of people never pay attention to the fact that learning often implies 'having been wrong'. Remember that next time someone tells you that *you did it wrong*.

Being wrong is not easy. Some people don't like being told. They get angry. Some people — especially people from certain countries and cultures — react in a sort of apologetic way: 'Oh, but I did it because…', 'How was I to know that…', 'It won't happen again.' I consider it my first task to attack that ingrained defence mechanism full-force.

- You did it with all the best intentions (trust).
- You are not an idiot (confidence).
- That's it. Stop whining. Loosen up and learn.

Invariably, at one time or another, I introduce the famous 'Four stages of competence'. None of my students had ever heard of them before. Actually, I only learned about them myself a couple of years ago from a friend who studied adult learning. It is astonishing because the concepts are so simple, general, and intuitive.

The two key ingredients are awareness and competence. Multiply these two and you get four states:

1. unconsciously incompetent
2. consciously incompetent
3. consciously competent
4. unconsciously competent

Think about the sequence of migrating from 1 to 4. When you do that, if you're a bit like me, you'll get a smile on your face. Such simple words, so elegantly wording what we all go through. To be able to get to the state of automatically handling something right, you have to learn how to do it, which you can only do if you understand that you couldn't do it (right) before. The key thing here

Being wrong is not easy.

Some people don't like being told.

They get angry.

is the planting of the seed: the realization that there is something not right.

Now apply the four states on a meta-level. You realize learning needs to go like this, so you start changing your response:

'Hey Joe, what's that? You're doing it wrong!'

'Eh?...But...hummmpf...OK.'

This in turn increases your social skills. People love being right, and like to be acknowledged. Now you have learned you were incompetent in handling your incompetence, you are starting to learn how to handle it. You're moving to the consciously competent state in successfully managing your incompetence. And after a while, you get to stage four. You never hesitate to do the right thing:

'Hey Joe, what's that? You're doing it wrong!'

'Wow, thanks man! I owe you one!'

And this, ambitious geophysicists, is the key skill you need to reach the top. Yes there are people who have reached some kind of top without it, but may I call your attention to the problem of the local maximum? Go for the global maximum. Be good at being wrong.

You are a geologist

Derik Kleibacker

I began my geoscience education in standard fashion, sketching line-drawings of outcrop bed geometries from road-cuts, scrubby creek bank-cuts, and the occasional quarry. Eventually the class moved on to exercises in stereographic aerial photography interpretation, again utilizing at its core simple lines on a page to represent geologic interpretation. The act of tracing lines on paper is almost insultingly literal, yet the geologic implications of the lines, the hypothesis that is laid bare, can be a thing of reasoned beauty.

Later I spent several seasons creating geologic maps from disparate pieces of spatial and temporal information, but looking back, I was just practicing how to meaningfully draw the lines. I was learning how to hypothesize within the constraints of the available evidence, guided by classical geologic reasoning. One summer I had a chance to work with 3D seismic data and things have never been the same. The amount of geologic information available in one place absolutely staggered me. I have been a proud seismic interpreter (a.k.a. geologist) ever since.

Philosophy and tips

Seismic reflection interpretation is one method of geologic interpretation. In order to get better at it you need to practice constructing and destructing testable geologic stories or hypotheses. Some may label this approach as 'model driven'; I suggest geoscientists are required to drive a testable model to be debunked, improved, or expanded. You will need to grasp all the options and weigh their merits before being able to meaningfully assign risk or resource to your geologic assessments.

The following are a set of simple tips for crafting an interpretation from seismic reflection data:

- Obtain and integrate with your interpretation every piece of geologic information available.
- Remember, the top of a geologic package must have a bottom.
- Always look for and map packages of geometries, usually called sequences.
- Avoid mapping just one surface at a time.

In order to get better at seismic reflection interpretation, you need to practice constructing and destructing testable geologic stories or hypotheses.

- As you pick seismic markers, think about what those surfaces represent geologically.
- Is that bright marker you were drawn to a downlap feature, an unconformity, or just a bright marker that you can map on a few lines?
- What are the implications of your current pick for the upcoming strike line? What should you see if you are right with your current model?
- Visualize what the map will look like before hitting the *Grid* button.
- When working with 2D data, interpret between the lines. What are you predicting to exist between data points?
- If you don't hold the working hypothesis in the forefront of your mind while interpreting then you are just an expensive autotracker; be a geologist.

The more familiar you become with interpreting while hypothesizing, the quicker you will be at calculating options and understanding as many geologic arguments fitting the data as possible. People may see this insight as a form of blind intuition, when in fact it is an attainable-through-practice science reasoning skill. Seek to habitually sharpen your geologic reasoning skills while interpreting. When done fluently, the results will be a clear and comprehensive understanding of the possible subsurface models, appropriately scaled to the available data quality and coverage.

Never forget to assess the confidence you have in the evidence that is feeding your hypothesis. With new data comes a recalibration of the geologic story, don't be afraid to change. It is distressing, but sometimes interpretations need to die.

List of contributors

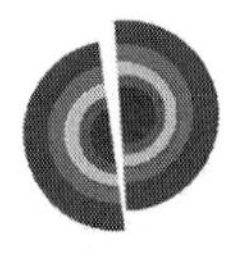

Eric Andersen is a senior geophysicist with Talisman Malaysia where he has a technical advisory role for Talisman's Southeast Asia Region (Malaysia, Indonesia, and Vietnam). He specializes in quantitative interpretation, seismic processing QC, and reservoir characterization, and has an interest in estimating geomechanical properties from seismic data. Eric received his BSc in mathematics in 1983 and a degree in geophysics in 1989 from the University of Alberta, Canada.

Per Avseth is a geophysical adviser at Odin Petroleum in Bergen, Norway, and a consultant to oil companies within the field of quantitative seismic interpretation and rock physics analysis. He is also an adjunct professor in applied geophysics at the Norwegian University of Science and Technology (NTNU) in Trondheim, Norway. Per received his MSc in applied petroleum geosciences from NTNU in 1993, and his PhD in geophysics from Stanford University, California, in 2000. He was the SEG Honorary Lecturer for Europe in 2009. Per is a co-author of the book *Quantitative Seismic Interpretation* (Cambridge University Press, 2005).

Art Barnes received a BS (1974) in physics from Denison University, an MS (1980) in geophysics from the University of Arizona, and a PhD (1990) in geophysics from Cornell University. His work experience includes seismic data acquisition, data processing, software development, and research. He is currently a geophysical researcher with Petronas in Malaysia. His research interests include seismic attributes, seismic processing, and pattern recognition applied to problems of seismic interpretation. He is a member of SEG, EAGE, and AAPG.

Evan Bianco is the Chief Scientific Officer at Agile Geoscience. He is a blogger, freelance geophysicist, entrepreneur, and knowledge-sharing aficionado. He has an MSc in geophysics from the University of Alberta and four years' experience as an industry consultant in Halifax, Nova Scotia. Evan's interests span a range of disciplines from time-lapse seismic in oil sands to geomodelling, seismic rock physics, and geothermal reservoir characterization. Evan tries to teach himself something new every day, and every so often, it proves useful. He can be reached at *evan@agilegeoscience.com*, or you can follow him on Twitter *@EvanBianco*.

Clare Bond is a structural geologist at the University of Aberdeen. She graduated from the University of Leeds and completed a PhD at Edinburgh University. Clare worked in geological conservation and policy roles in a range of fields before taking a research position co-hosted by the University of Glasgow and Midland Valley Exploration. She worked full-time for Midland Valley, working on consulting projects worldwide and leading their knowledge centre and R&D initiatives. Clare's research interests include uncertainty in seismic interpretation, and fault and fracture influence on sub-surface reservoirs. Having worked in both academia and industry she enjoys interdisciplinary research and has collaborated with social scientists, psychologists, and modellers.

Bert Bril is a co-founder and head of R&D at dGB Earth Sciences, *dgbes.com*. Bert holds an MSc in geophysics from Utrecht University. He began his career in 1988 as an acquisition geophysicist with Delft Geophysical, switched to software development in 1991, and worked for Jason Geosystems until 1992. He then worked at the TNO Institute of Applied Geoscience before co-founding dGB in 1995. He currently focuses on special projects and supporting his software team as an internal consultant.

José M Carcione was born in Buenos Aires, Argentina in 1953. He received the degree Licenciado in Ciencias Físicas from Buenos Aires University in 1978, the degree Dottore in Fisica from Milan University in 1984, and a PhD in geophysics from Tel-Aviv University in 1987. He has worked at the Comisión Nacional de Energía Atómica at Buenos Aires, and at Yacimientos Petrolíferos Fiscales, the national oil company of Argentina. Presently, he is a senior geophysicist at the Istituto Nazionale di Oceanografia e di Geofisica Sperimentale (OGS) in Trieste, Italy. In 2007, he received the Anstey award at the EAGE in London. José has published more than 200 journal articles on acoustic and electromagnetic numerical modelling, with applications to oil exploration and environmental geophysics. He is the author of the book *Wave fields in Real Media* (Elsevier Science, 2007) and co-author of *Arqueogeofísica* (Fundación de Historia Natural, 2006). He has been an editor of *Geophysics* since 1999.

Pavlo Cholach is a geophysicist with Abu Dhabi National Energy Company (TAQA) working on conventional oil assets in Alberta and Saskatchewan. Pavlo's previous seven-year experience with BP included geophysical support of a variety of conventional (structured carbonated and clastic gas reservoirs) and unconventional (tight gas, coal-bed methane, and oil sands) Canadian assets. Pavlo holds a PhD in geophysics from the University of Alberta, Canada, and a diploma with distinction in geophysics from Taras Shevchenko National University of Kyiv, Ukraine.

Alan J Cohen is Chief Geophysicist and Manager of Processing and Interpretation at RSI in Houston. He has over 30 years of experience in global oil and gas exploration. He has worked in operating units, in technical service organizations, and in research and development. He is regarded as an industry expert in rock physics of clastic and carbonate reservoirs, and in quantitative seismic interpretation. He was formerly Chief Geophysicist for Royal Dutch Shell in the western hemisphere. He can be reached at *alan.cohen@rocksolidimages.com*.

Why you care about Hashin–Shtrikman bounds 112

Alex Cullum has a BSc in geology and a PhD in palynology from the University of Aberystwyth, Wales. In 1996 he started work at IKU SINTEF in Trondheim, Norway, and since 1997 has worked in exploration for Statoil in Stavanger, Norway.

Seek out the biostrat 74

Paul de Groot is president of dGB Earth Sciences, *dgbes.com*. He worked 10 years for Shell where he served in various technical and management positions. Paul subsequently worked four years as a research geophysicist for TNO Institute of Applied Geosciences before co-founding dGB in 1995. He has authored many papers covering a wide range of geophysical topics and co-authored a patent on seismic object detection. Together with Fred Aminzadeh, Paul wrote a book on soft computing techniques in the oil industry. He holds MSc and PhD degrees in geophysics from Delft University of Technology in the Netherlands. Find him on LinkedIn at *ageo.co/IhwEUg*.

No more innovation at a snail's pace 48

Duncan Emsley graduated with a BSc from the University of Durham in 1984. He worked for processing contractors for several years before joining Phillips Petroleum in 1992. Continuing in the seismic processing vein, he worked with data from all sectors of the North Sea and northeast Atlantic. The merger of ConocoPhillips brought about moves to Scotland, Alaska, and Calgary, Canada, and a progression into rock physics and seismic attributes and their uses in the interpretation process.

Know your processing flow 38

Maitri Erwin is a geophysicist at an independent oil and gas company, where she currently specializes in the inversion and interpretation of deepwater data. She has previously worked as a 3D geospatial technology executive as well as a geoscientist and subsurface information technologist at an oil major. Maitri is an advisor to Project Gutenberg, the oldest producer of free electronic books. She is *@Maitri* on Twitter.

We need integrative innovation 102

Tooney Fink has been involved in the Canadian energy industry as a geophysicist for over 37 years, working for Gulf Canada Resources and successive companies which have since joined together to become ConocoPhillips Canada where he is now Supervisor of Geophysical Services. After graduating from the University of British Columbia in 1974, Tooney's career in the Canadian energy industry involved seismic field acquisition in the Mackenzie Delta and Parsons Lake, seismic processing, exploration in many of Canada's frontier basins, several international projects, and an endless number of projects in the Western Canada Sedimentary Basin. Tooney has been an active member of the CSEG Chief Geophysicists Forum since 2001, and helped compile *Practice Standard For Use Of Geophysical Data*. In 2008, Tooney was the CSEG General Chairman at the CSEG/CSPG/CWLS joint convention. Tooney also holds membership in AAPG, APEGA, NAPEGG, and SEG.

Remember the bootstrap 68

Victoria French is interested in all aspects of geosciences and even delves into the dark side of engineering when the opportunity presents itself. She holds BSc and MSc degrees from Baylor University and a PhD from the University of Oklahoma. Her favorite projects are integrated ones that incorporate geological, geophysical, and engineering inputs into models for reservoir characterization. Having held a number of technical and management roles in the last 17 years, she is currently Technical Director, Subsurface for TAQA NORTH. In this role, Victoria acts as technical advisor to TAQA NORTH's management team. She gives credit for her success to the incredible mentors that have inspired her and shaped her future: Emily Stoudt, Susan Longacre, Greg Hinterlong, and Kim Moore.

What I learned as a geophysicist wannabe 108

Taras Gerya is a professor at the Swiss Federal Institute of Technology (ETH-Zurich) working in the field of numerical modelling of geodynamic and planetary processes. He earned a PhD in petrology from Moscow State University in 1990 and a habilitation in geodynamics from ETH-Zurich in 2008. His recent research interests include modelling of subduction and collision processes, ridge-transform oceanic spreading patterns, intrusion emplacement into the crust, and core formation of terrestrial planets. He is the author of *Introduction to Numerical Geodynamic Modelling* (Cambridge University Press, 2009).

Calibrate your intuition 18

Bill Goodway obtained a BSc in geology from the University of London and an MSc in geophysics from the University of Calgary. Prior to 1985 Bill worked for various seismic contractors in the UK and Canada. Since 1985 Bill has been employed at PanCanadian and then EnCana in various capacities from geophysicist to Team Lead of the Seismic Analysis Group, to Advisor for Seismic Analysis within the Frontier and New Ventures Group, and subsequently in the Canadian Ventures and Gas Shales business unit. In 2010 he ended his career with EnCana to join Apache as Manager of Geophysics and Advisor Senior Staff in the Exploration and Production Technology group. Bill has received numerous CSEG Best Paper awards as well as the CSEG

Medal in 2008. He is a member of the CSEG, SEG, EAGE, and APEGA as well as the SEG Research Committee. In addition, Bill was elected Vice President and President of the CSEG for the 2002–04 term and in 2009 he was selected as the SEG's Honorary Lecturer for North America.

David Gray received a BSc in geophysics from the University of Western Ontario (1984) and a MMath in statistics from the University of Waterloo (1989). He worked in processing for Gulf, Seismic Data Processors, and Geo-X Systems and in reservoir and research for Veritas and CGGVeritas. He now works for Nexen, where he is a Senior Technical Advisor responsible for geophysical reservoir characterization in the Oil Sands group. David is a member of SEG, CSEG, EAGE, SPE, and APEGA. He has published and presented more than 100 papers, holds two patents, and is a registered Professional Geophysicist in Alberta.

Vladimir Grechka received his MSc degree (1984) in geophysical exploration from Novosibirsk State University, Russia and a PhD (1990) in geophysics from the Institute of Geophysics, Novosibirsk where he worked from 1984 to 1994 as a research scientist. He was a graduate student at the University of Texas at Dallas from 1994 to 1995, then joined the Colorado School of Mines where he was Associate Research Professor and co-leader of the Center for Wave Phenomena. From 2001 to 2012, Vladimir worked at Shell as a Senior Staff Geophysicist. Currently he is a Geoscience Consultant with Marathon Oil. Vladimir is the author of three books and more than 300 research papers devoted to various aspects of elastic wave propagation and seismic exploration. He is a member of SEG and EAGE. In 2009–2011, he served as the Editor-in-Chief of *Geophysics*. He received the East European Award from the European Geophysical Society in 1992 and the J Clarence Karcher Award from the SEG in 1997.

Matt Hall is the founder of Agile Geoscience. A sedimentologist who found geophysics later in his career, Matt has worked at Statoil in Stavanger, Norway, Landmark and ConocoPhillips in Calgary, Alberta, and is now running Agile from its world headquarters: an office conveniently located in his garden. He is passionate about communicating science and technology, and especially about putting specialist knowledge into the hands of anyone who needs it. Find Matt on Twitter as *@kwinkunks* or by email at *matt@agilegeoscience.com*.

Marian Hanna is President and Director of ION/GX Technology Canada. She has 23 years of diverse experience in the oil and gas industry. Marian started out as a seismic processing geophysicist with Amoco, moving into interpretation with an emphasis on reservoir characterization. Marian's experience includes many collaborative technical and business contributions to discoveries in international and domestic North American basins from onshore to deep water settings in all aspects of exploration, development and production, including business development/new ventures. Marian is a native of New Orleans, Louisiana.

Don Herron received a ScB in geological sciences from Brown University and an MS in geological sciences from the California Institute of Technology in 1973. He enjoyed a career as a seismic interpreter at Texaco from 1973–77, Gulf from 1977–1984, and Sohio/BP from 1984–2008. Since retirement in 2008 he has worked as an independent geophysical consultant for PGS and with several oil companies as a seismic interpretation instructor. He was co-instructor for the SEG course *Seismic Interpretation in the Exploration Domain* from 1995–2007. He was a member of the editorial board of *The Leading Edge* from 2002–07, its chairman from 2006–07, and is author of the bi-monthly *Interpreter Sam* column in *The Leading Edge*. He is also author of SEG Geophysical Monograph Series #15, *The Misadventures of Interpreter Sam*, and Geophysical Monograph Series #16, *First Steps in Seismic Interpretation*. He is an active member of SEG, AAPG, and Sigma Xi.

Chris Jackson completed his BSc (1998) and PhD (2002) at the University of Manchester, UK before working as a seismic interpreter for Norsk Hydro in Bergen, Norway. He left Norsk Hydro in 2004 to take an academic position at Imperial College, London, where he is now Reader in Basin Analysis.

Linn Margareth Johansen graduated from the University of Bergen with an MSc in marine geology and micropaleontology. She has worked at Statoil since then and is currently a team leader in biostratigraphy in Stavanger, Norway.

Chris Kent is from the UK and graduated from Camborne School of Mines. He has 17 years experience working in various places in Asia, West Africa, Saudi Arabia, North America, and Europe. He currently works as a geophysicist at Talisman Energy in Stavanger, Norway.

Derik Kleibacker is currently the Chief Geologist for ConocoPhillips Indonesia. He graduated with an MSc degree in geology from Oregon State University and has worked for ConocoPhillips in various exploration and development roles since 2002. Derik hates to admit it, but he loves geophysical interpretation as much as geological field mapping.

You are a geologist 116

John Logel is lead geophysicist for Talisman Energy in Stavanger, Norway, leading them in the application of value-adding geoscience technology to improve prospect quality and reduce risk. Prior to Talisman, John held several technical management and advisory positions with Anadarko Canada and Petro-Canada in Calgary, Alberta, and before that worked 19 years for Mobil on numerous assignments in Europe and North America. John teaches professional development courses for PetroSkills in basic geophysics, AVO, inversion, and attributes. John is a professional geophysicist and holds a BS and MS from the University of Iowa. He is a member of SEG, CSEG, APEGA, and AAPG.

Simplify everything 76

Dave Mackidd started his career at Esso in Calgary, Alberta, and enjoyed exploring the Canadian Arctic before moving to Canadian Hunter and, later, PanCanadian and EnCana. Throughout his career, he has been a prolific creator of geophysical tools and workflows, devising such marvels as the grid balance algorithm for 2D seismic lines. He is now a Geophysical Advisor at EnCana, pushing the boundaries of geophysics every day.

My geophysical toolbox, circa 1973 46
The evolution of workstation interpretation 80
The last fifteen years 86

Mostafa Naghizadeh received a BSc in mining engineering from the University of Kerman, Iran, and an MSc in geophysics from the University of Tehran in 2003. He received his PhD in geophysics from the University of Alberta in 2009. He worked as a postdoctoral researcher with CREWES at the University of Calgary from 2010–11. His interests are in seismic data reconstruction methods, sampling theory, and seismic imaging. He currently holds a postdoctoral fellow position with Signal Analysis and Imaging Group (SAIG) at the University of Alberta. In 2011, Mostafa received the J Clarence Karcher Award from SEG.

The magic of Fourier 88

Rachel Newrick obtained her BSc (geology; 1992) and BSc Honours (applied geophysics; 1993) at Victoria University of Wellington, New Zealand and her PhD (exploration seismology; 2004) at the University of Calgary. Since 1992 she has worked for BHP Petroleum in Melbourne, Warburg Dillon Read in London, Telecom NZ and Bank of NZ in Wellington, Occidental Petroleum in Houston, Exxon Mobil in Houston, Veritas DGC in Calgary, Nexen Inc in Calgary, and Nexen Petroleum UK in London. She is currently a senior geophysicist for Cairn Energy in Edinburgh working in frontier exploration. Rachel is the co-author of SEG Geophysical Monograph Series #13, *Fundamentals of Geophysical Interpretation* with Laurence Lines, has presented at a variety of conferences, and is presently the chairperson of the SEG Special Projects Committee.

Matteo Niccoli graduated from the University of Rome, Italy, with an honors degree in geology, and holds an MSc in geophysics from the University of Calgary. He worked for Canadian Natural Resources in Calgary, Alberta, and moved to Norway in the summer of 2012 where he is a senior geophysicist for DONG Energy. In his free time he does research and occasional consulting in geoscience visualization with his company MyCarta. On his blog he writes about exploration data enhancement and visualization, as well as image processing and its applications in geoscience, medical imaging, and forensics. He is a professional geophysicist of Alberta, and a member of AAPG, CSEG, and SEG. He can be contacted at *matteo@mycarta.ca* and is *@My_Carta* on Twitter.

Carl Reine graduated from the University of Alberta in 2000 with a BSc in geophysics. He worked a variety of conventional and heavy oil fields with Nexen Inc until 2006, when he attended the University of Leeds to complete a PhD. His research involved developing an algorithm for robustly measuring seismic attenuation from surface seismic data. Since 2010, Carl has worked for Nexen in the shale gas group, where he works on projects defining fracture/fault behaviour and reservoir characterization. He is an active member of CSEG, SEG, and EAGE, and is a professional member of APEGA.

Brian Romans is a sedimentary geologist and assistant professor in the Department of Geosciences at Virginia Tech. He graduated from SUNY Buffalo with a geology degree in 1997 and then worked as a geotech for small oil and gas companies in Buffalo, New York and Denver, Colorado for a few years. Brian received an MS in geology from Colorado School of Mines in 2003 and then headed to California where he earned a PhD in geological and environmental sciences from Stanford University in 2008. He worked as a research geologist for Chevron Energy Technology from 2008 to 2011 before joining the faculty at Virginia Tech. Brian's research on the patterns and controls of clastic sedimentation during and since graduate school have resulted in numerous papers, which you can access at *www.geos.vt.edu/people/romans*. Brian is *@clasticdetritus* on Twitter and writes the blog *Clastic Detritus* where he shares thoughts and photos about earth science.

Brian Russell started his career as an exploration geophysicist with Chevron in 1976, and worked for Chevron affiliates in both Calgary and Houston. He then worked for Teknica and Veritas in Calgary before co-founding Hampson-Russell in 1987 with Dan Hampson. Hampson-Russell develops and markets seismic inversion software which is used by oil and gas companies throughout the world. Since 2002, Hampson-Russell has been a fully owned subsidiary of Veritas and Brian is currently Vice President of Veritas Hampson-Russell. He is also an Adjunct Professor in the Department of Geology and Geophysics at the University of Calgary. Brian was President of the CSEG in 1991, received the CSEG Meritorious Service Award in 1995, the CSEG medal in 1999, and CSEG Honorary Membership in 2001. He served as chair of *The Leading Edge* editorial board in 1995, technical co-chair of the 1996 SEG annual meeting in Denver, and as President of SEG in 1998. In 1996, Brian and Dan Hampson were jointly awarded the SEG Enterprise Award, and in 2005 Brian received Life Membership from SEG. Brian holds a BSc in geophysics from the University of Saskatchewan, an MSc in geophysics from Durham University, UK, and a PhD in geophysics from the University of Calgary. He is a registered Professional Geophysicist in Alberta.

Mihaela Ryer has more than 17 years of experience in the oil and gas industry. She has worked for Marathon Oil, the ARIES Group, Prospectiuni (in Romania), and ConocoPhillips. Mihaela's research interests have been in the fields of seismic, sequence stratigraphy, depositional systems analysis and prediction. Most recently, she has employed a process-based approach to the prediction of lithofacies distribution and three-dimensional architecture of clastic systems through time and space, by using data-constrained stratigraphic forward-modelling technologies and tools. She is *@mihaela4021* on Twitter.

Marc Sbar got his PhD in earthquake seismology from Columbia University in 1972, and enjoyed a research and teaching career at Lamont-Doherty Earth Observatory and the University of Arizona until 1983. He then changed gears and spent 18 years as a geophysical specialist at BP, before moving to Phillips Petroleum in 2000, staying on at ConocoPhillips until 2009. Returning to academia, Marc recently moved to Tuscon, Arizona, impressing the next generation of University of Arizona geoscientists with the wonders of geophysics.

Rob Simm is a seismic interpreter with a special interest in rock physics, AVO, and seismic inversion technologies. Rob's early career (1985–1999) was spent with British independent oil and gas exploration companies Britoil, Tricentrol, and Enterprise Oil, working in both exploration and production. An interest in applying rock physics in prospect generation and field development led him to set up his own consultancy, Rock Physics Associates. His training course *The Essentials of Rock Physics for Seismic*

Amplitude Interpretation is recognized worldwide. Rob is the author of numerous papers as well as co-author of *3-D Seismic Interpretation* (Cambridge University Press, 2007). Since May 2010, Rob has had a staff position as Senior Geophysical Advisor at Agora Oil and Gas, a North Sea exploration company.

Sven Treitel grew up in Argentina and was educated at MIT where he graduated with a PhD in geophysics in 1958, before enjoying a long career at Amoco. Sven has published over 40 papers and is the recipient of numerous learned society awards, including the 1969 SEG Reginald Fessenden award, and in 1983 was awarded Honorary Membership of SEG. While his interests have been broad and varied, his main contribution to the field of geophysics has been to bridge the gap between signal processing theory and its application in exploration geophysics. He is the co-author of the definitive volumes *Geophysical Signal Analysis* (Prentice-Hall, 1980 & SEG, 2000) and *Digital Imaging and Deconvolution* (SEG, 2008). Although officially retired, Sven still lectures and consults widely.

Fereidoon Vasheghani received his PhD in geophysics in 2011, and MEng in petroleum engineering in 2006, both from the University of Calgary, Alberta. He is currently working as a geophysicist at ConocoPhillips. He is a member of SEG and SPE.

Index

J

K

L

M

N

Available from ***ageo.co/52geology*** and online bookstores worldwide.

ISBN 978-0-9879594-2-3

Discounts for orders of 10 copies or more.
Email ***hello@agilelibre.com*** for more information.

Printed in Great Britain
by Amazon.co.uk, Ltd.,
Marston Gate.